POF Simulation beyond Data Transmission

Summary of the 3rd International POF Modelling Workshop 2015

*We would like to thank all participants for their contributions,
the fruitful discussions, and enjoyable time we had.
We acknowledge the financial support of the
German Research Foundation (DFG).*

Christian-A. Bunge and Roman Kruglov (Ed.)

POF Simulation beyond Data Transmission

Summary of the 3rd International
POF Modelling Workshop 2015

© 2015 C.-A. Bunge, R. Kruglov
Herstellung und Verlag:
BoD – Books on Demand, Norderstedt
ISBN 978-3-7392-1499-3

Bibliografische Information der Deutschen Nationalbibliothek:
Die Deutsche Nationalbibliothek verzeichnet diese Publikation in der Deutschen Nationalbibliografie; detaillierte bibliografische Daten sind im Internet über http://dnb.d-nb.de abrufbar.

Contents

I. Introduction 3
    Some introductory remarks from the Editors 5

II. Light propagation in POF 7
    Molecular structure of optical polymers 9
    Modal dispersion in the frequency domain 23
    Temperature dependence of mode coupling 35
    Coupling losses 47
    Back scattering 57

III. Special types of POF 69
    Tubular fibres 71
    Active fibres 79

IV. Sensor applications 93
    Gamma Radiation Induced Effects in Perfluorinated POFs 95
    POF sensors in energy, oil and biotechnology areas 105
    Optical power monitor with SI-POFs 123

V. Author biographies 133

I

Introduction

Some introductory remarks from the editors

This book contains the contributions of the 3rd International POF Modelling Workshop on September 21, 2015. After a six year's pause since the previous workshop in Sydney, the POF Modelling Workshop returned to Nuremberg, POF-Application Center. It continues, however, the series of the scientific events, which are focused on different approaches to simulate polymer optical fibres (POF) and light propagation effects. Having a similar nature to glass multimode fibres POFs, however, require specific models because of their large numerical aperture, strong mode coupling and high attenuation. The scope of this year's issue still contained POF simulation and modelling, but encouraged contributions of applications other than communication systems.

During the last years emerged more and more a noticeable trend towards special applications at the International Conference on Plastic Optical Fibers. Since the POF-based Gigabit transmission systems have been coming to the market and the reported throughput of POF has been approaching the practical limit, a considerable number of the contributions is focused on sensor applications. That is why the scope had been widened to other topics. The workshop was even co-located with the TRIPOD meeting on POF sensors, which shared the first two talks joining both groups at the beginning. The keynote was connected with POF modelling in non-data transmission context.

The ten dedicated articles included in this book have been grouped in three thematic parts covering light propagation effects in POF, special types of POF and sensor applications. The reported advances in POF modelling, scientific problems being solved by the authors throughout the last years and new ideas related to POF applications surely benefited from participating in those workshops, which provide a discussion platform for the scientists and impact positively on the further development.

This year's workshop brought together researchers from four countries: Brazil, Germany, Serbia and Spain. The majority of the participants have participated in the previous modelling workshops as well.

The contributions were quite diverse and covered different topics ranging from polymer material modelling and measurement techniques to sensor applications and special fibre types such as tubular fibres. C.-A. Bunge from the HfT Leipzig presented a joint work with RWTH Aachen university on a molecular dynamic model to simulate the cooling behaviour of amorphous polymers. This work stood in the context of a novel POF production process for gradient-index polymer optical fibres without dopants. T. Becker from the POF Application Center in Nuremberg related on the impact of modal dispersion on the impulse and frequency response of POF. B. Drljača from the University of Kosovska Mitrovica in Serbia reported on the temperature dependence of mode coupling in SI-POF, which was simulated on the basis of Gloge's power-flow equation. M.A. Losada of the Universidad de Zaragoza, Spain, presented a method to assess coupling losses due to misalignments and fibre-parameter mismatches. M. Gehrke, also from the POF Application Center in Nuremberg, showed the simulation of backscattering in POF for a better understanding of OTDR measurements in these fibres. B. Lustermann from Nordhausen University of Applied Sciences, Germany, presented an interesting talk about ray-tracing modelling of tubular polymer optical fibres being a part of the optical-electrical conductor system. F. Jimenez from University of the Basque Country UPV/EHU, Spain, gave a presentation on the numerical simulation of light propagation and photon generation in POFs doped with fluorescent molecules.

Three presentations were focused on the practical implementation of the POF-based sensors: P. Stajanca of the BAM Federal Institute for Materials Research and Testing in Berlin, Germany, reported on the experimental investigation of radiation-induced attenuation in perfluorinated polymer optical fibres that can be used for the gamma-radiation monitoring. M.M. Werneck from the Universidade Federal do Rio de Janeiro, Brazil, gave an overview on POF sensors for the electric energy sector developed in Photonics and Instrumentation Laboratory. Finally reported R.M. Ribeiro from the Universidade Federal Fluminense, Brazil, on the implementation of an optical power monitor and its insensitivity to the modal distribution of the propagating light.

We can conclude that the given workshop was an attempt to involve the scientific groups into common discussion on the modelling of POF and its environment for POF sensors. This approach shall generate synergies between different research topics in a highly interdisciplinary field of science. We hope that this book contributes a little to this goal.

Christian-Alexander Bunge and Roman Kruglov
Leipzig/Nuremberg in December 2015

II

Light propagation in POF

- molecular structure and dynamics of optical polymers
- modal dispersion in the frequency domain
- temperature dependence of mode coupling
- coupling losses
- back scattering

Influence of the Cooling Speed on the Density of a Polymer: A Molecular-level Simulation Approach

C.-A. Bunge,[1,*] M. Beckers,[2] D. Grothe,[2] G. Seide,[2] T. Gries[2]

[1]*Institute for Communication Engineering, Hochschule für Telekommunikation Leipzig, Gustav-Freytag-Str. 43-45, D-04277 Leipzig, Germany.*

[2]*Institut für Textiltechnik, RWTH Aachen University, Otto-Blumenthal-Str. 1, D-52074 Aachen, Germany.*

*Corresponding author: bunge@hft-leipzig.de

In the context of a novel production process for gradient-index polymer optical fibers (GI-POF) without dopants, we present a molecular dynamic simulation in order to simulate the cooling behavior of amorphous polymers. The simulation uses the rotational isomeric state model and the statistical coil model. Intermolecular interactions are calculated based on the Lennard-Jones potential. A bounding volume hierarchy optimizes the processing time due to fast search of interacting chain elements.

1. Introduction

A novel, dopant-free and continuous production process for GI-POF has been developed to be more cost efficient and flexible [1,2]. The refractive-index profile is accomplished by a controlled density variation in the fiber. This density gradient arises by a rapid cooling of the molten polymer, which results a cooling gradient. To create the cooling gradient the fiber leaves the spinning nozzle into a temperature-controlled water quench. To adjust the retention time of the fiber the water quench has a height adjustable leading roller. By control of the duration and the temperature of the cooling treatment it is aimed to obtain a particular refractive-index profile [1-3].

We report on a Monte-Carlo model for the cooling of polymer based on the interaction of molecular chains. The goal of this simulation approach the description of the relation between cooling speed and the resulting density of the polymer. Once the density is known it can be related to a refractive-index.

2. Polymer models

There exist a whole range of different models to estimate real polymer chains. Each model has specific benefits and disadvantages. The following section gives a short overview over the probably most suitable polymer models and explains our choice of model. For a summary of the most suitable models please refer to Fig. 1.

- The **Edwards-Tube Model** describes the dynamic behavior of entangled polymers. The complex interactions of the polymers are averaged and the movement is restricted to a simple tubus [4].

- In the **Rotational Isomeric-State Model (RIS)** the length and angle of the bond to each chain element is constant. The model offers three different rotation angles that represent the three energetically favorable states [5].

- The **Random-Coil Model** assumes the polymer chains to entangle themselves in an unordered manner. The chains distribute statistical around the center of mass [6].

- In the **Meander Model** the polymer chains fold themselves into an energetic favorable conformation. The polymer chains arrange themselves into so called meander cubes, clusters of folded chains [7,8].

For the final modeling of the polymer's cooling behavior we chose to implement the RIS model combined with the Random-Coil model. The decision was based considering the complexity in the computation of the models. The chosen model offers a decent compromise of accuracy and computation times.

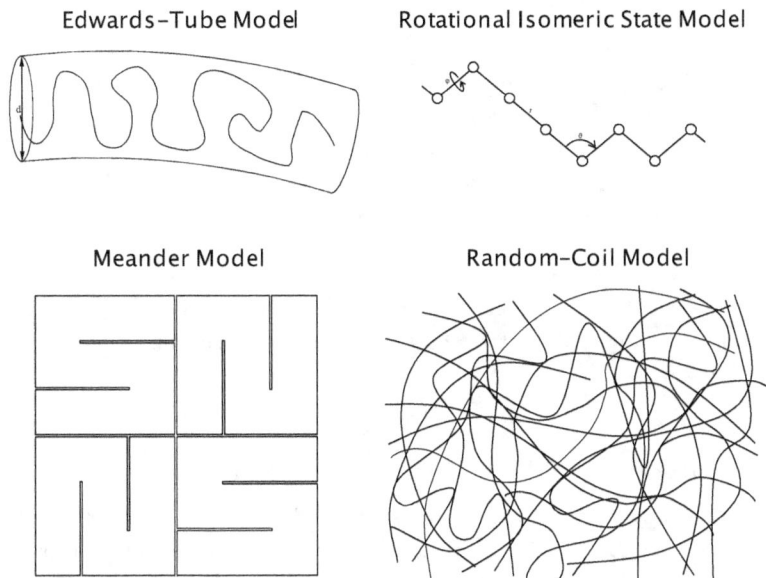

Fig. 1: Overview of suitable polymer models.

3. Implementation aspects

In the following we describe the implementation of the modified rotational isomeric state model. We give an overview about the internal data structures and used algorithms. The initialization of the variables is described as well as the implementation of the intermolecular interactions.

3.A. Chains and chain elements

To describe the polymer chains the simulation uses the RIS model. Fig. 2 shows a schematic of a polymer-chain representation in the RIS model. Each chain is represented by an individual object. A chain object contains a list with all its chain elements, the current position vectors both of the center of mass and the first chain element and the current velocity of the chain. Each chain element for itself is represented by an individual object and contains the absolute position vector, the position vector relative to the previous element in the chain and the angle of torsion to the next element.

The chosen object oriented approach ensures an easy and intuitive understanding of the implementation of the model. Each element can be viewed and calculated separately. Future changes to how interactions between chains or chain elements are treated, e. g. the addition of intra-molecular interactions, can easily be implemented.

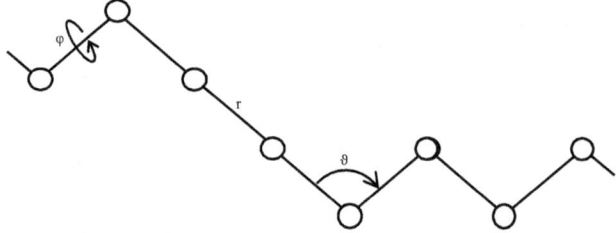

Fig. 2: Schematic of a polymer chain based of the RIS model, φ is the angle of torsion, ϑ is the bonding angle and r the distance between the chain elements [5].

3.B. Initial variables

The simulation uses a set of initial variables specified by the user. The initial variables describe the general conditions of the simulation. Some variables indicate the initial state the model shall assume, others are responsible to define to accuracy and computational effort of simulation. Table 1 shows the specified initial variables.

Table 1: List of variables used in the polymer model.

variable	unit	description
N	–	number of polymer chains used in the simulation
n	–	number of elements each polymer chain consists of
M	$\frac{km}{mol}$	molar mass of each chain element
ϑ	rad	bonding angle between two chain elements
r	m	distance between each chain element
T	–	total number of time steps the simulation computes
Δt	s	time step
$Temp$	K	initial temperature used for chain generation
RB	m	defines the area in which the polymer chains are created
a	m	cut-off radius
ϵ	J	depth of the potential well
σ	m	distance at which the Lennard-Jones potential approaches zero

The initial variables are divided into two groups. N, n, M, ϑ,r, $Temp$ and RB are used to generate the polymer chains and their initial positions randomly. The initial velocities of the chains follow a Maxwell-Boltzmann distribution. Fig. 3 shows an example for initial velocities of 20,000 chains with 50 elements each.

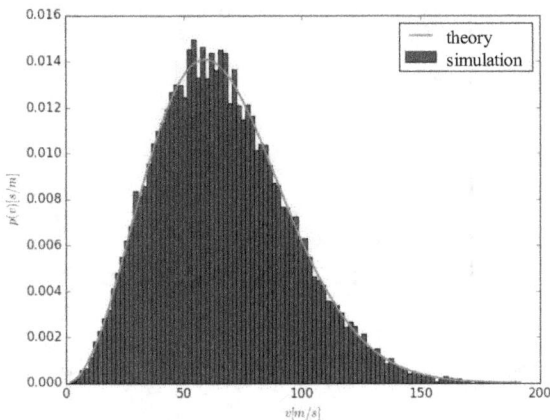

Fig. 3: Exemplary probability distribution of initial velocities with $N = 20000$, $n = 50$.

The variables T, Δt, a, σ and ϵ set the conditions of the simulation during runtime. T and Δt describe the number and step size of the time steps during the simulation. The smaller the step size Δt is set the higher is the simulation accuracy potentially. Naturally, a higher accuracy goes with a smaller simulated period of time or higher computation time. The variables a, σ and ϵ are used to calculate the interactions of the chain elements (see Sec. 3.C).

3.C. *Implementation of the molecular interactions*

The interactions between the chain elements are calculated from the Lennard-Jones potential, which provides an estimation of both van-der-Waals attraction and Pauli repulsion:

$$V(r) = 4\epsilon \left[\left(\frac{\sigma}{r}\right)^{12} - \left(\frac{\sigma}{r}\right)^{6} \right]. \tag{1}$$

The variable ϵ describes the depth of the potential well, while σ defines the distance between the two interacting elements where the potential equals zero. ϵ and σ are characteristics to the chosen chain element. r is the distance between the two interacting chain elements. Fig. 4 shows the course of the Lennard-Jones potential over the distance.

The derivation of the Lennard-Jones potential $\frac{dV(r)}{dr}$ describes the force that is interacting between the two chain elements [9,10].

$$\frac{dV(r)}{dr} = -F(r) = 24\epsilon \left[\frac{\sigma^6}{r^7} - \frac{\sigma^{12}}{r^{13}} \right] \tag{2}$$

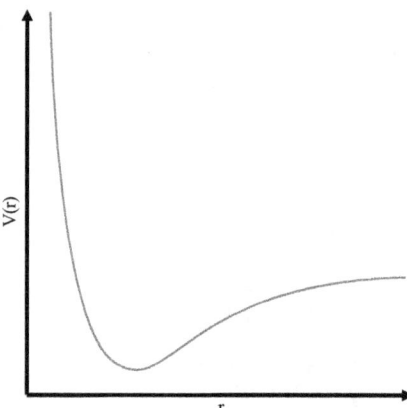

Fig. 4: Dependence of the Lennard-Jones potential $V(r)$ on the distance r.

As can be seen in Fig. 2, the Lennard-Jones potential and with it the force between the interacting elements decreases with greater distance. At a particular distance is the force small enough to be neglected. This distance is defined as cut-off radius a, which is specified at the beginning of the model. Calculating only the interactions of elements within the cut-off radius speeds up the simulation significantly. Only the nearby elements have to be considered [11].

With all forces calculated, the motion of the rigid chains can be done. The translation and rotation motions of the chain are calculated by standard equations of motion:

$$\vec{r}_l(t + \Delta t) = \vec{r}_l(t) + \vec{v}_l(t) \cdot \Delta t + \frac{\vec{F}_l(t)}{2m_i} \cdot \Delta t^2 \tag{3}$$

$$\vec{v}_l(t + \Delta t) = \vec{v}_l(t) + \frac{\vec{F}_l(t)}{m_i} \cdot \Delta t \tag{4}$$

$$\vec{\varphi}(t + \Delta t) = \vec{\varphi}(t) + \vec{\omega}_l(t) \cdot \Delta t + \frac{1}{2} \frac{\partial \vec{\omega}_l(t)}{\partial t} \cdot \Delta t^2 \tag{5}$$

$$\vec{\omega}_l(t + \Delta t) = \vec{\omega}_l(t) + \frac{\partial \vec{\omega}_l(t)}{\partial t} \cdot \Delta t \tag{6}$$

Here, $\vec{r}_l$ describes the positional vector and $\vec{v}_l$ the velocity of the l-th chain. m_l is its mass, while $\vec{\varphi}_l$ describes the angle and $\vec{\omega}_l$ the angular velocity of the l-th chain. The angular acceleration $\frac{\partial \vec{\omega}_l}{\partial t}$ is calculated as

$$\frac{\partial \vec{\omega}_l(t)}{\partial t} = \mathbf{I}^{-1} \cdot \vec{M}_{tot}, \tag{7}$$

where $\mathbf{I}$ describes the moment of inertia and $\vec{M}_{tot}$ the cumulated moment each force applies to the center of mass of a chain.

3.D. Bounding-volume hierarchy

The Lennard-Jones potential decreases with greater distance between the interacting elements. The cutoff-radius limits the number of chain elements that can interact with each element based on their distance to each other. To efficiently find all significantly interacting elements a so-called binding-volume hierarchy (BVH) is introduced.

A BVH is a data structure based on a binary search tree. A binary search tree is a recursive structure. Each node of the tree stores data and has two sub-trees. Each sub-tree leads to another node. A binary search tree is sorted and allows for fast iteration and therefor fast searches. An example of a general binary search tree is shown in Fig. 5. In the example numbers are sorted along the tree. Lower numbers are moved to the left, higher numbers to the right sub-tree. [12,13]

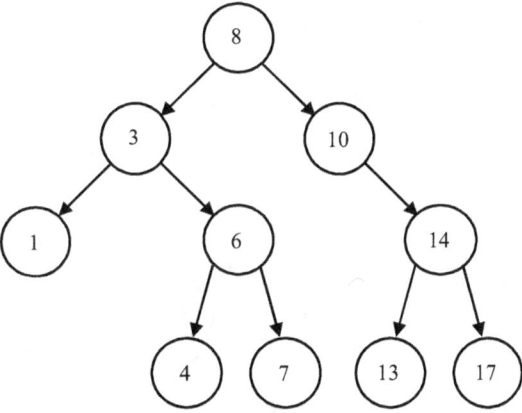

Fig. 5: Example of a binary search tree.

A BVH splits the complete simulation volume with every sub-tree. Fig. 6 shows an example of a BVH and an example of the recursive steps the algorithm takes to find interacting elements. With every recursive stage in the BVH tree the potential volume gets split and refined. To find all interacting elements of one viewed chain element the algorithm takes the position vector of the viewed element and searches the binary search tree for all elements within the cutoff-radius. At every step of recursion the positional vectors of the viewed element is compared to the current potential volumes on the right and left sub-trees. If the viewed element is within the potential volume, the specific sub-tree will be followed. If the viewed element is not within the potential volume of the sub-tree, the sub-tree will be left. Due to the nature of the binary search tree the algorithm is very fast [14].

In comparison to the initial approach of testing the distance of every element pair, the new implemented BVH potentially offers a linear dependence of processing time. A single search in a binary search tree has an average complexity of $\mathcal{O}(\log_2 n)$ with n describing the absolute

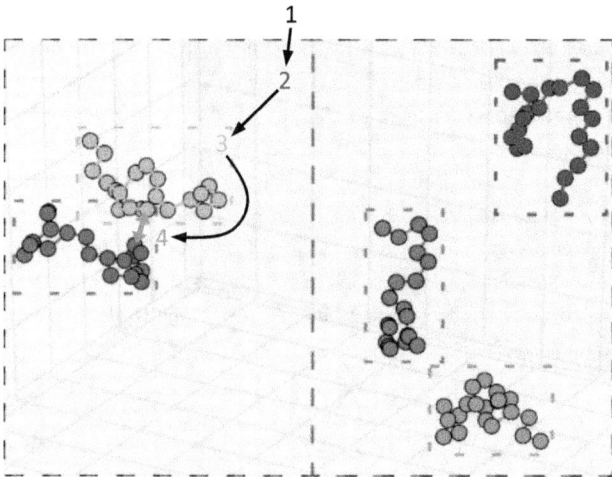

Fig. 6: Structure of a bounding volume hierarchy: Steps 1 to 4 illustrate the recursive search for interacting chain elements.

number of chain elements. The initial approach of testing the distance of every element pair has an average complexity of $\mathcal{O}(n)$. The potential gain of processing speed using the BVH is $\mathcal{O}(n)/\mathcal{O}(\log_2 n)$, which is approximately linear. Taken time measurements approve the linear correlation [13].

Further improvements of the BVH can be accomplished. Currently the BVH has to be generated each time step. The BVH is a static data structure and the chains constantly move out of the generated volumes. To compensate for the constant movement of the chains a so called "fat factor" can be introduced. The fat-factor increases the size of the volumes by a specific factor. The chains do not immediately move out of the bigger volumes. Therefore the BVH does not need to be re-generated every time step.

3.E. Implementation of evaluation aspects

Important parameters to evaluate are the density and the temperature of the system. Since the density is defined as mass per volume it can be determined by counting the chain elements in a specific volume. The average density $\bar{\rho}_i$ in the volume V_i of the system calculates from the mean average of multiple densities of small random volumes.

$$\bar{\rho}_i = \frac{N_i}{V_i} \cdot M \qquad (8)$$

with N_i being the number of polymer-chain elements in the the volume V_i.

The average temperature of the system is calculated from the velocities of the chains using the Maxwell-Boltzmann distribution $v = \sqrt{2k_B T/m}$:

$$\overline{T} = \frac{1}{N} \sum_{i=1}^{N} \frac{v_i^2 3 \cdot m_i}{2k_B}, \tag{9}$$

where v_i describes the velocity of the i-th chain. m_i is the mass of the chain and k_B is the Boltzmann constant. The average temperature $\overline{T}$ is the mean average from each chain in the system.

The cooling behavior can be simulated using the Berendsen thermostat. The idea is to put the system into a cooling bath. The average temperature of the system converges on to the temperature of the cooling bath:

$$\frac{\partial T(t)}{\partial t} = \frac{T_0 - T(t)}{\tau}. \tag{10}$$

Here, T_0 is the temperature of the cooling bath and τ a time constant, which adjusts the strength of the thermal coupling. The Berendsen thermostat can be implemented directly into the equations of motion. A change of temperature is equivalent to the change of velocity of the chains. [9,15]

4. Results

As a test of our model and the implementation into the simulation several simulations with different initial parameters have been done. The following sections give an overview of the results of the simulations and how those results support the chosen model and its implementation.

4.A. Volume expansion for different initial temperatures

As described in Sec. 3.A, the initial velocities of the polymer chains are generated using the initial temperature specified prior to the simulation. Different initial temperatures lead to different expansion behavior. In the first test case the same simulation is done at several initial temperatures and the expansion behavior is observed during the course of the simulation. The polymer chains are expected to have higher velocities at higher initial temperatures. The simulation is performed with 3,000 polymer chains and 300 chain elements each. Calculations are carried out over 100 time steps. The compared initial temperatures are 1 K, 293 K (room temperature) and 400 K. Fig. 7 shows the expansion at each case.

As expected the simulation with the lowest initial temperature also has the slowest expansion. Due to the low velocities of the chains there is almost no motion in the system during the simulation. The higher the temperature gets the higher is the velocity of the chains. This leads to a higher expansion of the system during the simulation.

4.B. Influence of the cut-off radius on the accuracy

For the sake of better calculation performance, a cut-off radius ensures that only nearby chain elements, which mainly interact with each other, will be taken into account. The elements at a higher distance can be neglected without large errors. To confirm this assumption a

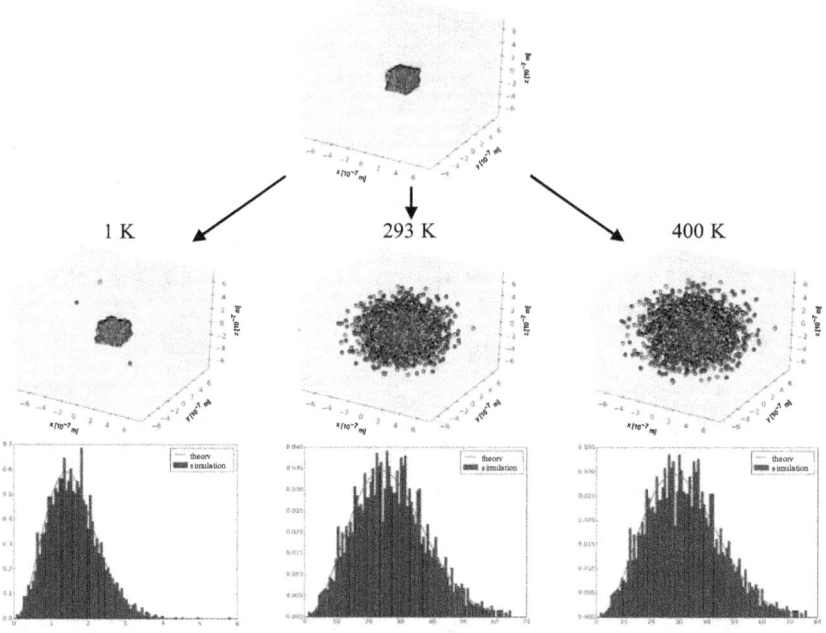

Fig. 7: Evolution of volume expansion from the same initial state at different initial temperatures.

sample simulation is firstly performed with a cut-off radius of and afterwards without any cut-off radius. In the second case, in every time step the interactions between all elements are calculated. The calculation is performed on 500 polymer chains and 500 chain elements each. The initial temperature is set to 293 K and the calculations are done over 50 time steps. The expansion at the end of the simulation is depicted in Fig. 7.

Comparing the expansion of both simulations after 50 time steps shows that calculating every possible interaction leads to very similar results than considering only the nearby chain elements. The assumption that the interactions with higher distance can be neglected is valid.

4.C. Influence of time-step size on accuracy

Finer time steps offer potentially higher accuracy at the cost of longer processing times. Coarse time steps, however, speed up the simulation, but are potentially inaccurate. The choice of the right step size of the time increments is a compromise between these two extremes.

In our test cases, we tested several time steps to elaborate what inaccuracies can appear if the time steps are too coarse. The main issue with too coarse time steps is a higher gain of velocity of chains at high repulsion. Due to the high time steps the force of the repulsion

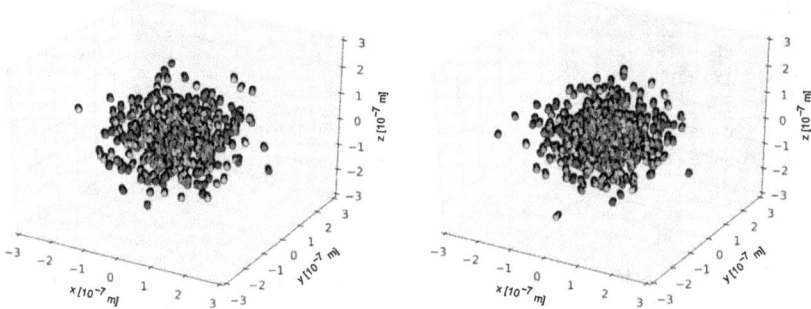

Fig. 8: Comparison of calculated volume expansion with and without cutoff-radius.

between two interacting chains affects the chains for too long. A long acceleration and high gain of velocity is the result, which leads the chains to expand way faster than they should be. Typically with time steps of about 1 fs to 10 fs the simulation is precise enough [9]. However, coarser time steps can often be tolerated because of missing high repulsion. In those cases time steps up to 10 ps were sufficient. A possible way of using this to our advantage is to detect high repulsive interactions during the simulation and altering the time steps on the fly.

5. Conclusion and outlook

We have presented a simulation framework to predict the molecular dynamics within polymers at different velocities and temperatures. One important next step is the implementation of boundary conditions. At the moment no boundary conditions are implemented. The simulation volume is not restricted and once a chain leaves the cutoff radius of every other chain behind it cannot be retrieved. This leads to a continuous expansion of the system volume and a continuous decrease of the average density [9].

The polymer chains are based on the rotational isomeric state model. The polymer chains are currently rigid and the simulation offers only intermolecular interactions. The implementation of intramolecular interactions can increase the accuracy of the results as the polymer chains can be calculated more realistically.

With faster the calculations more chains and chain elements can be simulated. With the implementation of a bounding volume hierarchy came a significant decrease of computation time. Further optimizations of the algorithms and the use of multiprocessing can decrease the processing time even further.

The main goal of the simulation is to simulate the cooling behavior of polymers to get a specific refractive index gradient as a function of the density. The results show that the core concepts of our model lead to reasonable results. In the next steps, the simulation has to advance to a point where the interplay between cooling speed and material density can be described.

Acknowledgements

The research was partly funded by Deutsche Bundesstiftung Umwelt (DBU) and the Bundesministerium für Bildung of Forschung (BMBF) of Germany.

References

1. M. Beckers, T. Schlüter, T. Vad, T. Gries, C.-A. Bunge, "An overview on fabrication methods for polymer optical fibers," Polymer International, Vol. 64, No. 1, pp. 25-36, Jan. 2015, DOI: 10.1002/pi.4805.
2. M. Beckers, T. Schlüter, T. Gries, C.-A. Bunge, "An innovative extrusion technology for the melt spinning of polymer optical fibers used for light transmission over short distances," International Fiber Journal, Vol. 28, No. 2, pp. 8-10, Feb. 2014.
3. M. Beckers, W. Steinmann, N. Holt, T. Vad, G. Seide, and T. Gries, Schmelzspinnverfahren für die Herstellung von über den Querschnitt variierbaren Fasern (GI-Profil) und ihre Verwendung, insbesondere optische Polymerfasern und im speziellen auf Basis von Polymethylmethacrylat (PMMA), Polystyrol (PS) oder Polycarbonat (PC) sowie anderen Commodity Polymeren. German patent 10 2013 009 169.1.
4. S.F. Edwards and T.A. Vilgis, "The tube model theory of rubber elasticity," Rep. Prog. Phys. 51 (1988), 243-297.
5. W.L. Mattice and U.W. Suter, "Conformational Theory of Large Molecules. The Rotational Isomeric State Model in Macromolecular Systems," Wiley, New York (1994).
6. U. Eisele, "Introduction to polymer physics," Berlin, New York: Springer Verlag (1990).
7. W.R. Pechhold, T. Gross, H.P. Grossmann, H.G. Zachmann, H.-G. Kilian, and R. Hosemann, "Meander model of amorphous polymers," Colloid and Polymer Science 260 (1982), 378-393.
8. W. Pechhold, "Meander Model of Amorphous Polymers," Makromol. Chem. Suppl 6 (1984), 163-194.
9. R. Hentschke, E.M. Aydt, B. Fodi, and E. Stöckelmann, "Molekulares Modellieren mit Kraftfeldern," Vorlesungsskript, Bergische Universität Wuppertal (2004).
10. M. Rubinstein, R.H. Colby, "Polymer physics," OUP Oxford (2003).
11. M. Kotelyanskii and D.N. Theodorou, "Simulation methods for polymers," CRC Press (2004).
12. K. Mehlhorn and P. Sanders, "Algorithms and data structures: The basic toolbox," Springer Science & Business Media (2008).
13. T.H. Cormen, C. Leiserson, R. Rivest, and C. Stein C, "Introduction to algorithms," MIT press (2009).
14. J.T. Klosowski, M. Held, J.S. Mitchell, H. Sowizral, K. Zikan, "Efficient collision detection using bounding volume hierarchies of k-DOPs," Visualization and Computer Graphics,

IEEE Transactions 4 (1998), 21-36.
15. P.H. Hünenberger, "Thermostat algorithms for molecular dynamics simulations," Advanced computer simulation, Springer Berlin Heidelberg (2005), 105-149.

Influence of Modal Dispersion on Impulse and Frequency Response of Step-Index Polymer Optical Fiber

T. Becker,[1,*] S. Loquai,[1] H. Poisel,[1] O. Ziemann,[1] M. Luber,[1] B. Schmauss[2]

[1]*Polymer Optical Fiber Application Center,*
Technische Hochschule Nürnberg, 90489 Nürnberg, Germany.

[2]*Institute of Microwaves and Photonics (LHFT),*
Friedrich-Alexander-Universität Erlangen-Nürnberg, 91054 Erlangen, Germany.

*Corresponding author: thomas.becker@th-nuernberg.de.

This paper investigates the impact of modal dispersion on both impulse response and the resulting frequency response. In a simplified manner modal dispersion is often believed to cause a pulse broadening of rectangular shape. However, the directional characteristic of the light source and a non-linear relationship between a mode's propagation angle and its transit time lead to a deviation from this shape. As a result, the phase response shows non-linear intervals.

1. Introduction

Optical length or strain measurement techniques often rely on a linear relationship between the length of an optical fiber and the phase shift that is caused by the required transit time [1–4]. From a system-theoretical point of view, the fiber can be treated as a transmission system with a corresponding impulse and frequency response that describes how the amplitude and phase of a modulated input signal are affected by the transmission. In order to achieve a linear relationship between the length of an optical fiber and the transit time for a modulated signal, the phase of the frequency response (also known as the phase response) has to be a linear function of the modulation frequency. This paper shows the impact of the shape of an impulse response caused solely by modal dispersion on the phase response.

2. Theoretical background

As mentioned in the introduction, the pulse broadening in an optical fiber prevents a linear relationship between its length and the phase shift of a modulated signal with which the fiber is excited. One of the major effects that are responsible for the deformation of the impulse response is modal dispersion. In an optical multimode fiber, the distribution of the electromagnetic field can take various forms which can be represented as modes or, in a simpler approach, as rays. These modes can be subdivided into different groups with different propagation properties [5]. However the length of the path of a ray inside a fiber as shown in Fig. 1 only depends on its angle α relative to the fiber's axis:

$$L_{ray} = \frac{L_{fiber}}{\cos(\alpha)} \qquad (1)$$

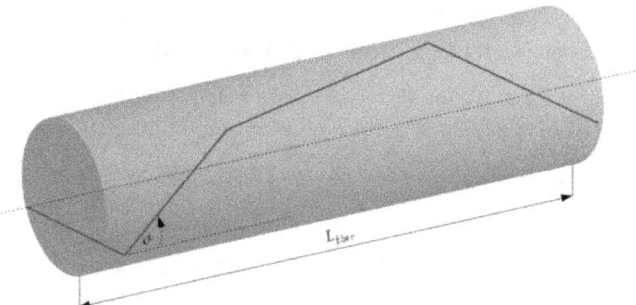

Fig. 1: Ray path length

2.A. Simulation of an impulse response

In a step-index fiber the time-delay d caused by a ray is proportional to its length:

$$d = \frac{n_{core} L_{ray}}{c_0}, \qquad (2)$$

with n_{core} being the refractive index of the core and c_0 being the speed of light in vacuum. Therefore the impulse response of a multimode step-index polymer optical fiber (SI-POF) with n rays can be described as a set of weighted Dirac-impulses $\delta(t)$ with different delays d_k, where each Dirac-impulse represents one ray:

$$g(t) = \sum_{k=0}^{n} \frac{1}{n} \delta(t - d_k) \qquad (3)$$

Since we do not consider absorption or any other kind of losses, the energy of the exciting Dirac-impulse is shared among the resulting Dirac-Impulses. A light source with a uniform power distribution over the solid angle is assumed. Therefore each Dirac-impulse is weighted with the factor n^{-1}. Examples for both $\delta(t)$ and $g(t)$ are shown in Fig. 2 on the next page. While the given expression for $g(t)$ is technically correct, it is unsuitable for the illustration of the actual pulse broadening. For this purpose another form of impulse-response $gs[m]$

is introduced. $gs[m]$ can be obtained from $g(t)$ by discretizing the time into equally defined intervals Δt with m being the index of the interval starting at 0. The energy of all rays arriving per interval is summed up to the total energy and the power of an interval can be calculated according to its length:

$$P_{interval}[m] = \frac{E_{interval}[m]}{\Delta t} \qquad (4)$$

An example for $gs[m]$ is shown in Fig. 2c.

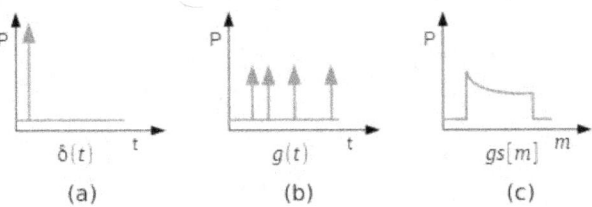

Fig. 2: Different interpretations of an Impulse response

2.B. Retrieving the phase and amplitude response of the fiber

If the fiber is considered to be a transmission system, its transfer function can be obtained by transferring the impulse response to the frequency domain. This is often done with the discrete Fourier Transform (DFT). However the DFT creates a transfer function for periodic functions in the time domain [6]. Since the impulse response is aperiodic, the calculated transfer function would contain an error. This error could be reduced by padding zeros at the end of the impulse response but it can never be fully suppressed.

An alternative algorithm which expects an aperiodic function is the discrete-time Fourier transform (DTFT) [7]. The DTFT creates a continuous transfer function:

$$F(\omega) = \sum_{m=-\infty}^{\infty} gs[m]e^{-i\omega m \Delta t} \qquad (5)$$

The calculated frequency response depends on the chosen value for Δt. Since the discretization of the time was only introduced to illustrate the pulse broadening, its effect on the resulting frequency response is undesirable. Instead of transferring $gs[m]$ with Eq. 5, it is possible to treat each ray of the simulation as a single transfer function:

$$F_{ray}(\omega) = \frac{1}{n}e^{-i\omega d_k} \qquad (6)$$

The transmission system of the fiber is the parallel arrangement of the transmission systems of all rays:

$$F(\omega) = \sum_{k=0}^{n} \frac{1}{n} e^{-i\omega d_k} \qquad (7)$$

With Eq. 7 the transmission characteristics of the fiber can be described independent of Δt. The corresponding amplitude and phase responses are the absolute value and the argument of the transfer function:

$$R_a(\omega) = |F(\omega)| \qquad (8)$$

$$R_p(\omega) = \arg(F(\omega)) \qquad (9)$$

The amplitude response describes how the amplitude of the modulated signal depends on the frequency after it has passed the fiber. The phase response shows how a modulated signal of a specific frequency is delayed.

2.C. Ideal phase response

As mentioned before the elongation measured by optical strain-gages is often believed to have a linear dependency on the modulation frequency. Since we want to calculate the error that is caused by modal dispersion, we have to determine the phase response for an ideal system, which is free from dispersion and on which this principle of measurement is based on.

The phase response is the ratio between the length of the fiber and the period length of the modulated signal[1] λ_{mod} multiplied by 2π. If we express the period length as

$$\lambda_{mod} = \frac{c_0}{fn_{core}} = \frac{c_0 2\pi}{\omega n_{core}}, \qquad (10)$$

the ideal phase response can be written as

$$R_{pi}(\omega) = \frac{L_{fiber} 2\pi}{\lambda_{mod}} = \frac{L_{fiber} \omega n_{core}}{c_0}. \qquad (11)$$

3. Simulations

So far we have shown how to obtain the phase response of a fiber based on the optical path lengths of the rays that are propagating in the fiber. For this paper we extract the required optical path lengths from ray tracing simulations created by *Synopsys LightTools*. We define a

---
[1] The period length of the modulated signal is not the wavelength of the light.

straight SI-POF with the length $L_{fiber} = 10\,\text{m}$ and the refractive indices $n_{core} = 1.495$ in the core and $n_{clad} = 1.42$ in the cladding. The diameter of the core is $d_{core} = 980\,\mu\text{m}$. The light source is emitting 2.5 million rays from a plane circular area of the same size as the fiber's core. It is placed on the fiber's axis and 0.01 mm in front of the fiber. The angular power distribution is uniform over the half space that is facing the fiber's end surface. Therefore it can be assumed that every possible mode in the fiber is excited.

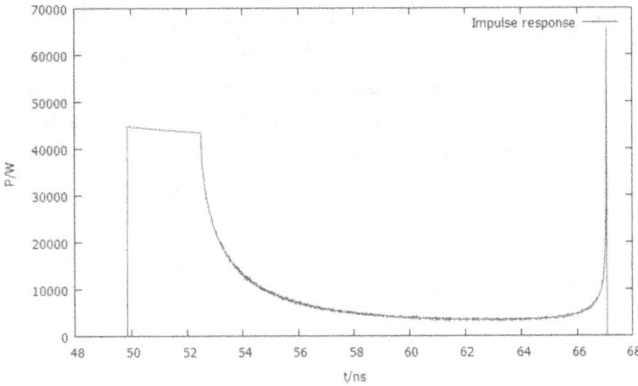

Fig. 3: Impulse response of 10 m SI-POF

Figure 3 shows the impulse response of the fiber that is created in the way described in the previous chapters. For the simulation a total energy of the exciting Dirac impulse of $E_{dirac} = 1\,\text{mWs}$ is defined. 546702 of the 2.5 million simulated rays reach the end of the fiber. The remaining rays are lost because they are not coupled into the fiber. Since every ray represents an equal amount of the total energy, about one fifth of E_{dirac} is spread over the width of the impulse response. The shape of the impulse response shows some interesting details which are easier to understand if we have a look at the corresponding modal distribution of the impulse response (Fig. 4), which shows the energy distribution over α. The critical angle for the total internal reflection in our fiber is

$$\alpha_{max} = \arccos\left(\frac{n_{clad}}{n_{core}}\right) = \arccos\left(\frac{1.42}{1.495}\right) = 18.23° \tag{12}$$

All rays with $\alpha < \alpha_{max}$ are guided no matter if they are meridional or skew rays [8]. In the modal distribution these rays represent the area between 0° and 18.23°. While the graph seems to be linear in that area, it actually has a sinus dependency on α. Since the time a ray needs to travel through a fiber depends on α, guided modes build the first section of the impulse response. All rays with $\alpha > \alpha_{max}$ are leaky modes and can be subdivided into refracted and tunneling modes. Refracted modes are not totally reflected at the core-cladding-interface and are therefore not part of the simulation. As a consequence, all rays in Fig. 4 with $\alpha > \alpha_{max}$ are tunneling modes. Tunneling modes do fulfill the critical angle for the total internal reflection and are therefore treated as guided modes in the simulation. However due to the curved core-cladding-interface the frustrated total internal reflection is causing a significant attenuation for these modes which is not considered by the simulation.

For the explanation of the part of the impulse response that is caused by tunneling modes we have to consider two effects. The first is the decreasing progression of the curve for angles beyond $18.23°$. If a ray with an angle $\alpha > \alpha_{max}$ is a refracted ray (and therefore not part of this discussion) or a tunneling ray depends on its starting position and the angle between the projection of the ray's path onto the fiber's end surface and the surface normal of the core-cladding-interface at the point of the reflection. However in a simplified manner one could say that the larger α, the less the probability that it is a tunneling mode.

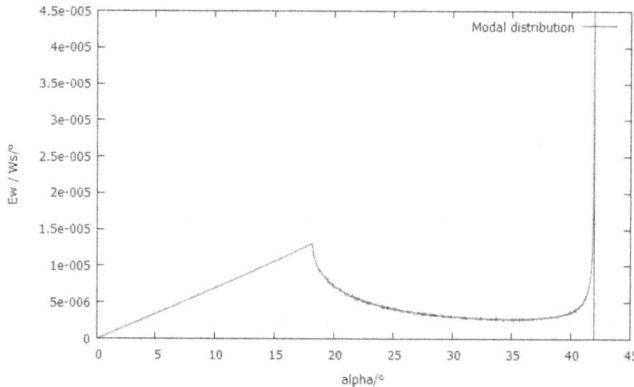

Fig. 4: Modal distribution of 10 m SI-POF

The second interesting detail of the impulse response is the power peak at the end. While the uniform power distribution over the solid angle corresponds to a sinusoidal relation between the power and the angle α_0[2] outside the fiber, the power distribution changes according to Snell's law when the rays enter the fiber. The modal distribution is the derivative of the energy of the exciting pulse with respect to α inside the fiber, or α_0 outside the fiber. Therefore we can write:

$$\frac{\mathrm{d}E}{\mathrm{d}\alpha_0} \propto \sin(\alpha_0) \tag{13}$$

---

[2] α_0 is the angle between a ray and the fiber's axis before it enters the fiber.

If we assume that the light source is surrounded by a material with the refractive index $n_0 = 1$ we can express the modal distribution inside the fiber as:

$$\begin{aligned}\frac{\mathrm{d}E}{\mathrm{d}\alpha} &\propto \frac{\mathrm{d}E}{\mathrm{d}\alpha_0}\frac{\mathrm{d}\alpha_0}{\mathrm{d}\alpha} \\ &\propto \sin(\alpha_0)\frac{\mathrm{d}\arcsin\left(n_{core}\sin(\alpha)\right)}{\mathrm{d}\alpha} \\ &\propto \frac{\cos(\alpha)\sin(\alpha)n_{core}^2}{\sqrt{1-\sin^2(\alpha)n_{core}^2}}\end{aligned} \quad (14)$$

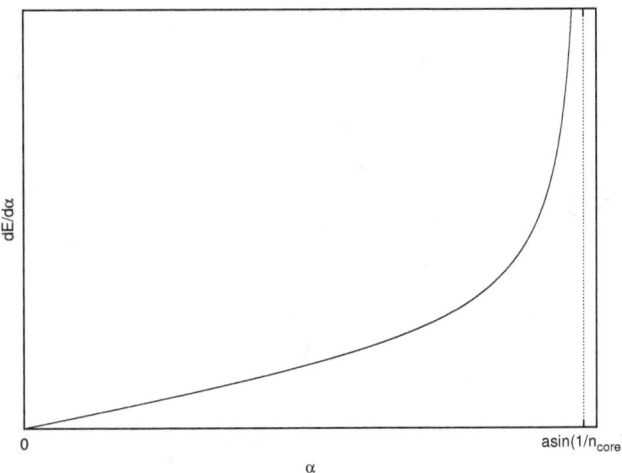

Fig. 5: Modal distribution after fiber entry

The resulting modal distribution after the fiber entry is depicted in Fig. 5. The power that is distributed between $0°$ and $90°$ outside the fiber is now compressed between $0°$ and $\arcsin(1/n_{core}) = \arcsin(1/1.495) = 41.98°$ and most of the power is shifted towards the maximum angle which explains the power peak at the end of the modal distribution and the impulse response respectively.

4. Comparison of the simulated phase response and the linear approximation

Figure 6 on the next page shows both, the phase response that is extracted from the simulation and the ideal phase response that is calculated according to Eq. 11 on page 26 for a fiber length of 10 m. In this plot the simulated phase response seems to be linear and its slope differs only slightly from the ideal phase response. However its exact shape is hidden by the scale of the diagram. Therefore the difference between both responses is shown in Fig. 7. From this depiction it becomes clear that the difference is not linear but shows an oscillating behavior. For the consequences of the phase error e_p for optical strain-gages it is important to calculate the resulting length error e_l from which a length measurement would suffer:

$$e_l = e_p \frac{c_0}{360° n_{core} f_{mod}} \qquad (15)$$

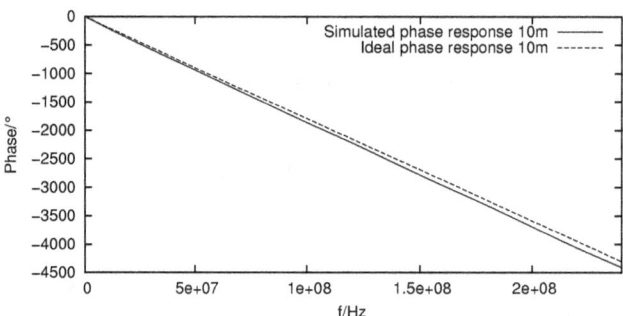

Fig. 6: Phase response of the simulation and the ideal theory

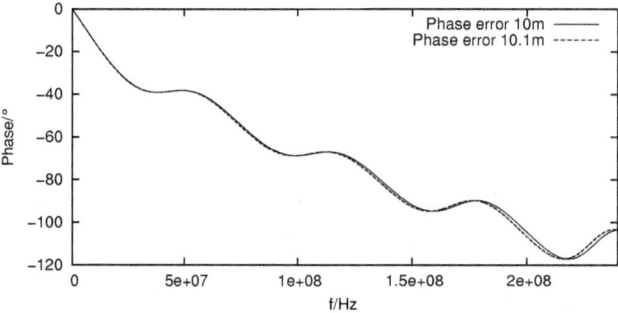

Fig. 7: Absolute phase error

The calculated length error e_l is plotted in Fig. 8 on the facing page up to $f_{mod} = 240\,\text{MHz}$. As an example we choose the frequency $f_{mod} = 100\,\text{MHz}$. According to Fig. 8 the error is 0.38 m. If the length of a 10 m fiber would be determined using a phase measurement at this frequency, its length would be measured to be 10.38 m. The plot shows a minimum error of about 0.24 m which, at a fiber length of 10 m, corresponds to a minimum error of 2.4%.

While the calculated error is relatively large, one has to keep in mind that optical strain sensors are usually used to detect length changes and not the absolute length of a fiber to which the error in depiction 8 would correspond to. Therefore we create a second simulation with the same parameters as the first one but with a changed fiber length of $L_{fiber2} = 10.1$ m. The resulting phase and length errors are depicted in Figs. 7 on the preceding page and 8 together with the simulated errors for the 10 m fiber.

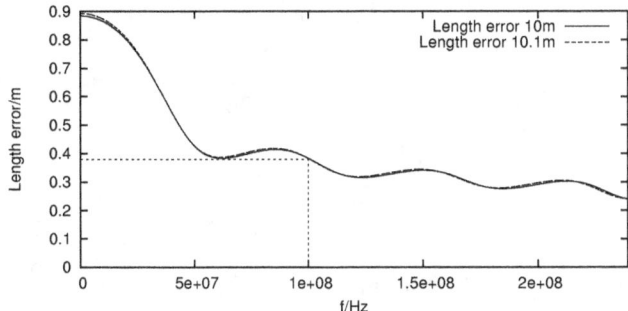

Fig. 8: Absolute length error

The measured length change for a fiber that is stretched from 10 m to 10.1 m can be determined from the difference of the simulated phase responses for both fibers:

$$\Delta l = \frac{R_{p2}(f_{mod}) - R_{p1}(f_{mod})}{360°} \cdot \frac{c_0}{n_{core} f_{mod}} \qquad (16)$$

Figure 9 on the next page shows the the calculated length change. Despite the fact that the minimum absolute length errors of the simulations are rather large, the calculated difference oscillates around the actual strain of 100 mm. Areas in which the calculated strain is larger than 100 mm are colored grey. The black areas mark a calculated strain that is less than 100 mm. It can be seen from the plot that it is theoretically possible to choose frequencies where no error is expected at all. In the plotted range this is the case for 38 MHz, 49 MHz, 99 MHz, 113 MHz, 158 MHz, 177 MHz and 217.5 MHz.

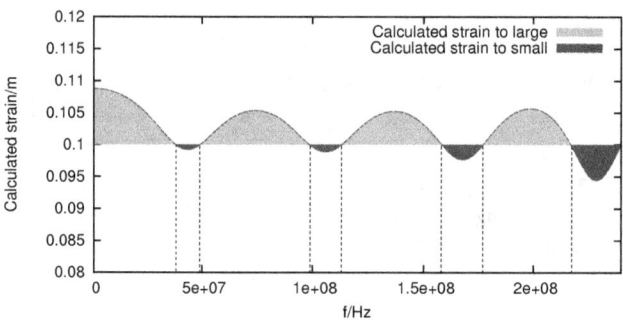

Fig. 9: Calculated strain

5. Conclusions

We investigated the influence of modal dispersion in SI-POF and the consequences for optical strain sensors that rely on the phase detection of a modulated signal. Ray tracing simulations were set up to retrieve the impulse responses of a SI-POF at different lengths. The impulse responses were transferred to the frequency domain via the DTFT and the resulting phase responses were examined. From these phase responses we calculated the modal dispersion induced error that affects absolute length measurements as well as strain measurements. The proposed approach can be utilized to increase the accuracy of optical strain sensors.

The strain measurement was simulated for a fiber of a length of 10 m and a strain of 100 mm. For this specific case the occurring error of the phase measurement was calculated depending on the modulation frequency. However for real applications it is important to know the measured error for any possible strain. Therefore a next step could be the calculation of the measurement error with respect to both, modulation frequency and actual strain. Since this requires a continuous spectrum of impulse responses depending on the strain, the chosen approach that is based on ray tracing simulations is not suitable. An alternative approach would be the analytical derivation of a fiber's impulse response depending on the strain.

6. Acknowledgements

The work described in this paper was supported by the grant Optika[2] from the Bayerisches Staatsministerium für Bildung und Kultus, Wissenschaft und Kunst. The authors gratefully acknowledge the support of the graduate study group Fiber Optic Transmission and Sensing (FiTS).

References

1. H. Poisel, "POF strain sensor using phase measurement techniques," in *Proceedings of SPIE Conference Smart Sensor Phenomena, Technology, Networks, and Systems, 6933-35*, San Diego, March 2008.
2. G. Durana, M. Kirchhof, M. Luber, I. S. de Ocáriz, H. Poisel, J. Zubia, G. Aldabaldetreku, and C. Vázquez, "Design and evaluation of a novel fiber optical elongation sensor," in *Proceedings of 17th International Conference, POF'08*, Santa Clara (USA), August 2008.
3. M. Luber, H. Poisel, S. Loquai, C. Neuner, A. Bachmann, O. Ziemann, and E. Hartl, "POF strain sensor using phase measurement techniques," in *Proceedings of 16th International Conference on POFs and Applications, POF'07*, Turin (Italy), September 2007, pp. 29–32.
4. M. Luber, H. Poisel, O. Ziemann, J. Diez, G. Durana, and M. Schukar, "Erfahrungen mit dem praktischen Einsatz eines faseroptischen Dehnungssensors," in *Proceedings of 19th International Scientific Conference Mittweida, IWKM 2008*, Mittweida, November 2008, pp. 26–30.
5. B. E. A. Saleh and M. C. Teich, *Grundlagen der Photonik*, 2nd ed. Boston: WILEY-VCH, 2008.
6. M. W. Wong, *Discrete Fourier Analysis*. Basel: Birkhäuser Verlag, 2011.
7. O. Föllinger, *Laplace-, Fourier- und z-Transformation*. VDE-Verlag, 2011.
8. A. W. Snyder and J. D. Love, *Optical Waveguide Theory*. Kluwer Academic Publishers Massachusetts, 2000.

Temperature Dependence of Mode Coupling in Low-NA Plastic Optical Fibers

S. Savović,[1,2,*] M. S. Kovačević,[1,3] J. S. Bajić,[4] D. Z. Stupar,[4]
A. Djordjevich,[2] M. Živanov,[4] B. Drljača,[5] A. Simović,[1] K. Oh[3]

[1] Faculty of Science, University of Kragujevac, R. Domanovića 12, Kragujevac, Serbia.

[2] City University of Hong Kong, 83 Tat Chee Avenue, Kowloon, Hong Kong, China.

[3] Institute of Physics and Applied Physics, Department of Physics, Yonsei University 134 Sinchon-dong, Seodaemun-gu, Seoul, South Korea.

[4] University of Novi Sad, Faculty of Technical Sciences, Serbia.

[5] Faculty of Science, University of Kosovska Mitrovica, Lole Ribara 29, Kosovska Mitrovica, Serbia.

*Corresponding author: savovic@kg.ac.rs

Using the power flow equation and experimental measurements, investigated in this article is the state of mode coupling in low NA (0.3) step-index (SI) plastic optical fibers under varied temperature. Numerical results obtained using the power flow equation agrees well with experimental measurements. It is found that elevated temperatures of low-NA SI plastic optical fibers strengthened mode coupling. These properties remained after a year, with the fiber being subjected to environmental temperature variations of more than 35 K. These thermally induced changes of the fiber properties are attributed to the increased intrinsic perturbation effects in the PMMA material at higher temperatures.

1. Introduction

The importance of plastic optical fibers (POFs) has grown tremendously over past decades. POFs are highly promising transmission media for short-range applications including in LANs, multi-node bus networks, sensors, power delivery systems, and light guides (as in toys, entertainment and medical devices). The attractiveness of POFs is chiefly due to their low cost, flexibility and ease of handling and interconnecting. However, their relatively large attenuation and modal dispersion limit the link length and transmission rate. The main types of POFs, their manufacturing and possible present and expected future applications have been reported

[1]. Much effort has been made in recent decades to mitigate the effect of limited bandwidth of SI-POFs [1,2]. Temperature variation as an avenue for additional improvements has been indicated by simulation [3] by showing that modal dispersion decreases as temperature is increased, leading to a bandwidth improvement. One should mention here that in reference [3] the simulations that concerned POFs were based on measurements for bulk PMMA and this is only a coarse approximation.

Ambient temperature can often affect structure of the POF material [4,5]. Intrinsic perturbation effects may also change with temperature [3,5] (these perturbations are by irregularities introduced predominantly during the fiber manufacturing process that resulted in microscopic bends, irregularity of the core-cladding boundary [6], and inconsistency of the refractive index). In spite of all fibers undergoing some degradation of optical transmission when exposed to heating, fiber-optic systems are nevertheless implemented in warm environments particularly as optical fiber sensors [5,7] and the influence of temperature may be considerable. Changes optical fibers undergo with temperature fluctuation have been investigated [3,5,8,9].

Transmission characteristics of multimode SI optical fibers depend heavily upon the differential mode attenuation and rate of mode coupling. The latter represents power transfer from lower to higher-order modes caused by the above-mentioned intrinsic perturbation effects. Mode coupling reduces modal dispersion, leading to a bandwidth improvement in local area networks (LANs) [10]. On the other hand, it increases the amount of power radiated in fiber curves or bends [11–13], significantly changing the output-field properties and degrading the beam quality. These consequences are difficult to predict intuitively and have a particular importance for power delivery and sensory systems [10,14].

In order to determine the fiber length L_c where an equilibrium mode distribution (EMD) is achieved, one can analyze the change of the output angular power distribution as a function of fiber length for different launch angles [14]. Coupling length L_c is the fiber length after which all output angular power distributions take the disk-form regardless of the incidence angle. The shorter the length L_c, the earlier the bandwidth would switch from the functional dependence of z^{-1} to $z^{-1/2}$ (slower bandwidth decrease).

Output angular power distribution in the near and far fields of an optical fiber end has been studied extensively. Work has been reported using geometric optics (ray approximation) to investigate mode coupling and predict output-field patterns [16,17]. By employing the power flow equation [15, 18–21], these patterns have been predicted as a function of the launch conditions and fiber length. A key prerequisite for achieving this is the knowledge of the rate of mode coupling expressed in the form of the coupling coefficient D [18–20], which has been shown to correctly predict coupling effects observed in practice (e. g. [15]).

In this work, we use the Savović-Djordjevich method [22] for determining the mode coupling coefficient D in a low-NA POF at temperatures ranging from 293.15 K to 353.15 K. This enabled us to solve the power flow equation numerically in order to examine the state of mode coupling in the fiber being analyzed at different temperatures. As a result, the coupling length L_c at which the EMD is achieved and the fiber length z_s required for achieving the steady-state mode distribution (SSD) are obtained for different fiber temperatures. These numerical results are verified experimentally.

2. Power flow equation

Gloge's power flow equation is [18]:

$$\frac{\partial P(\theta,z)}{\partial z} = -\alpha'(\theta)P(\theta,z) + \frac{D}{\theta}\frac{\partial}{\partial \theta}\left(\theta\frac{\partial P(\theta,z)}{\partial \theta}\right), \quad (1)$$

where $P(\theta,z)$ is the angular power distribution, z is distance from the input end of the fiber, θ is the propagation angle with respect to the core axis, D is the coupling coefficient assumed constant [10,13–15,19,20,23] and $\alpha'(\theta)$ is the modal attenuation. The boundary conditions are $P(c,z) = 0$, where θ_c is the critical angle of the fiber, and $D(\frac{\partial P}{\partial \theta}) = 0$ at $\theta = 0$. Condition P(c,z)=0 implies that modes with infinitely high loss do not carry power. Condition $D(\frac{\partial P}{\partial \theta}) = 0$ at $\theta = 0$ indicates that the coupling is limited to the modes propagating with 0. Except near cutoff, the attenuation remains uniform $\alpha'(\theta) = \alpha'_0$ throughout the region of guided modes $0 \leq \theta \leq \theta_c$ [20,21] (it appears in the solution as the multiplication factor $exp(-\alpha'_0 z)$ that also does not depend on θ). Therefore, $\alpha'(\theta)$ need not be accounted for when solving Eq. (1) for mode coupling and this equation reduces to [15]:

$$\frac{\partial P(\theta,z)}{\partial z} = \frac{D}{\theta}\frac{\partial P(\theta,z)}{\partial \theta} + D\frac{\partial^2 P(\theta,z)}{\partial \theta^2} \quad (2)$$

In order to obtain numerical solution of the power flow equation (2) we have used the explicit finite-difference method (EFDM) employed in our earlier works [13,15].

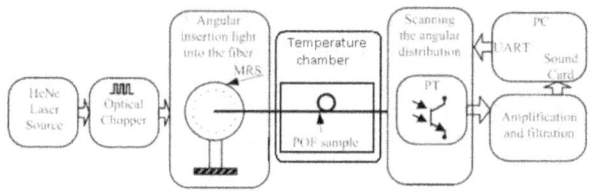

Fig. 1: Block diagram of the experimental setup for measuring the far-field intensity pattern (MRS – manual rotation stage, PT – phototransistor, PC – personal computer).

3. Experimental method

In this section, we present the experimental set-up and methodology used to obtain the far-field intensity in the MH4001 ESKA Mitsubishi Rayon fiber (MH fiber). The fiber has a core diameter $d_{core} = 980\,\mu$m and clad diameter $d_{clad} = 1$ mm, numerical aperture $NA = 0.3$, core refractive index $n = 1.49$ and critical angle $\theta_c = 11.7°$ measured inside the fiber and $\theta_c = 17.6°$ measured in air. The MH fiber has 0.15 dB/m of nominal attenuation. The number of modes in this step-index multimode plastic optical fiber, for $\lambda = 633$ nm, is $N = 2\pi^2 a^2 NA^2/\lambda^2 \approx 1.06 \times 10^6$, where a is radius of the fiber core. This large number of modes may be represented by a continuum as required for application of equation (2). The ends of the sample fibers have been polished carefully for reduced imperfections and associated light diffraction.

Fig. 1 shows schematics of the experimental setup for measuring the far-field intensity profile at different temperatures. The light source was a helium neon laser (HRR005 by Thorlabs) with maximum power of 0.5 mW at 633 nm, modulated by a chopper at $f = 700$ Hz. The input fiber end was mounted on a rotation stage (MRSPRM1/M by Thorlabs) in order to achieve an angled launch. Using a phototransistor (BPW17, 520–950 nm range), the far-field output intensity from the other fiber-end was profiled in 3D by a two-motor scanning setup.

For the temperature variation, a chamber with adjustable temperature has been used. The temperature has been also measured with an external high precision digital thermometer. The heating process consisted of increasing the temperature of the chamber by 20 K every 30 minutes, from 293.15 K to 353.15 K.

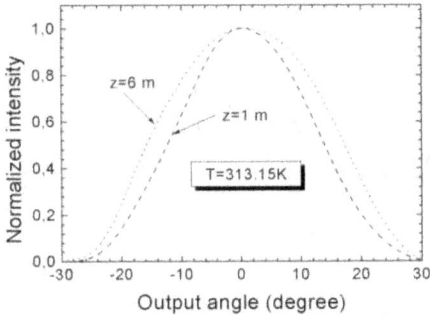

Fig. 2: Normalized experimental output power distribution at two different locations along the MH4001 plastic optical fiber for centrally launched ($\theta_0 = 0°$) Gaussian beam, at $z = 1$ m (dashed line) and $z = 6$ m (dotted line) at temperature T=313.15 K.

4. Results and discussion

In our recently published work [24], coupling coefficient D for this fiber was calculated as $D = 1.62 \times 10^3 \frac{\text{rad}^2}{\text{m}}$ at ambient temperature of T=293.15 K. From $D = \sigma_2^2 - \sigma_1^2 / 2(z_2 - z_1)$ [22], the coupling coefficient D was calculated as $D = 3.41 \times 10^3 \frac{\text{rad}^2}{\text{m}}$ for $\sigma_1 = 12°$ and $\sigma_2 = 16°$ being the two standard deviations of the experimental output angular power distribution at fiber lengths $z_1 = 1$ m and $z_2 = 6$ m (Fig. 2), at T=313.5 K. The values of the mode coupling coefficient D for the analyzed fiber at different temperatures are summarized in Table 1, respectively, to facilitate easier comparisons.

Starting with the fiber length of 15 m, this length was shortened by 0.25 m in successive steps until the final length of just 0.75 m remained. In each step, the launch angle was swept in 5° increments from 0° to 15° and for each such sub-step the far-field output pattern was recorded (by the photo detector in the two-motor scanning setup). The fiber placed inside the temperature-controlled chamber was coiled on a reel with diameter in excess of 0.5 m to minimize the influence of curvature on mode coupling.

Fig. 3 shows experimental results for the normalized output power distribution for different fiber lengths at $T = 313.15$ K. We show results for four different input angles $\theta = 0°$, 5°, 10° and 15° (measured in air). In Fig. 4, our numerical solution of the power flow equation is presented by showing the evolution of the normalized output power distribution with fiber length at temperature $T = 313.15$ K. A good agreement between the numerical and experimental results can be observed. Radiation patterns in the short fiber (z=1 m) in Figs. 3(a) and 4(a) indicate that the coupling is stronger for the low-order modes: their distributions have shifted towards $\theta = 0°$. Coupling of higher-order modes can be observed after longer fiber lengths (Figs. 3(b) and 4(b)). It is not until the fiber's coupling length L_c of 3 m that all the mode-distributions have shifted their mid-points to zero degrees (from the initial value of

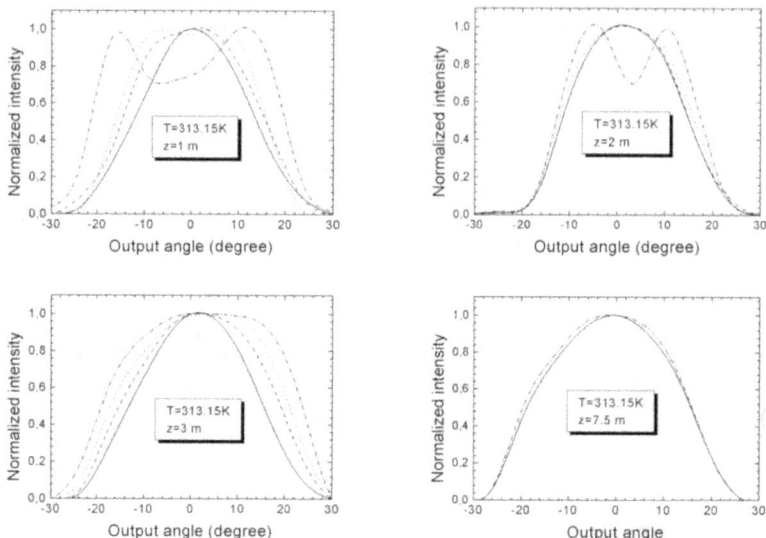

Fig. 3: Normalized **experimental** output angular power distribution at different locations along the MH4001 plastic optical fiber at $T = 313.15$ K for four Gaussian input angles $\theta_0 = 0°$ (solid line), $5°$ (dashed line), $10°$ (dotted line) and $15°$ (dash-dotted line) at: (a) $z = 1$ m (solid line); (b) $z = 2$ m (dashed line); (c) $z = 3$ m (dotted line) and (d) $z = 7.5$ m (dash-dotted line).

0 at the input fiber end), producing the EMD in Figs. 3(c) and 4(c). The coupling continues further along the fiber beyond the L_c mark until all distributions' widths equalize and SSD is reached at length z_s in Figs. 3(d) and 4(d): $z_s = 7.5$ m. The maximum estimated relative error of determining the coupling coefficient $\Delta D/D$, coupling length $\Delta L_c/L_c$ and length at which a SSD is achieved $\Delta z_s/z_s$ is approximately 10%.

It can be seen from Table 1 that with increasing fiber temperature there is an increase of the coupling coefficient D, meaning that intrinsic perturbation effects in the fiber are stronger. With increasing fiber temperature the width of the output angular power distribution measured at a fixed fiber length increases too (Fig. 5). Consequently, the lengths L_c and z_s (Fig. 6) shorten.

We have obtained that with increasing temperature, there is a very small change of numerical aperture of the fiber (Fig. 7, left). Namely, measured numerical aperture of MH4001fiber varies between NA =0.37 and 0.38, at temperatures between 293.15 K and 353.15 K, respectively. Thus, the influence of the change of the numerical aperture on temperature dependence of the coupling constant D can be excluded. One should mention here that measured numeri-

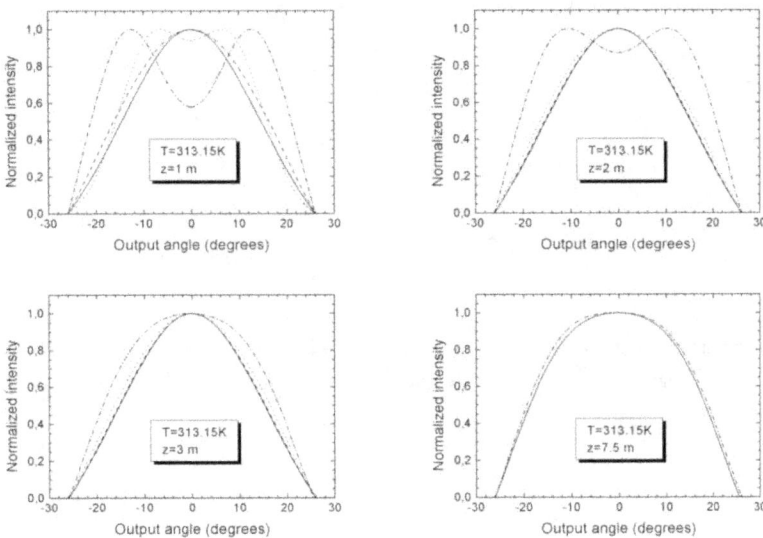

Fig. 4: Normalized **simulated** output angular power distribution at different locations along the MH4001 plastic optical fiber at $T = 313.15$ K for four Gaussian input angles $\theta_0 = 0°$ (solid line), $5°$ (dashed line), $10°$ (dotted line) and $15°$ (dash-dotted line) at: (a) $z = 1$ m (solid line); (b) $z = 2$ m (dashed line); (c) $z = 3$ m (dotted line) and (d) $z = 7.5$ m (dash-dotted line).

Table 1: Fiber temperature T, coupling coefficient D, coupling length L_c for achieving the EMD and length z_s for achieving the SSD.

Temperature T [K]	D [rad^2/m]	L_c [m]	z_s [m]
293.15	1.62×10^{-3}	4.0	10.0
313.15	3.41×10^{-3}	3.0	7.5
333.15	4.53×10^{-3}	2.25	5.5
353.15	5.82×10^{-3}	1.5	4.0

cal aperture of NA =0.37 at 293.15 K for MH4001 fiber differs from its theoretical numerical aperture of NA=0.3, due to the fact that this fiber is a double cladding fiber and not typical SI-POF [1].

We have also obtained that with increasing fiber temperature, there is a slight increase

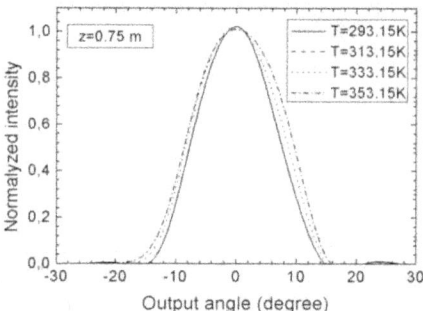

Fig. 5: Normalized experimental output angular power distribution along the MH4001 plastic optical fiber of length $z = 0.75\,\text{m}$ at different temperatures for centrally ($\theta = 0°$) launched Gaussian beam.

of mode dependent attenuation coefficient (Fig. 7, right). This is in contrast with results reported by Chen et al. [8] for the MH4001 plastic optical fiber with higher numerical aperture of NA=0.5. They have obtained that there is a decrease of power loss with increasing fiber temperature.

One can see from Fig. 8 that the modal attenuation at a given temperature could be considered approximately constant with the propagation angle in the range $\theta = 0°$ to $\theta \approx 18°$, which enabled us to numerically solve the power flow equation in its simpler form (2) instead of solving the power flow equation (1).

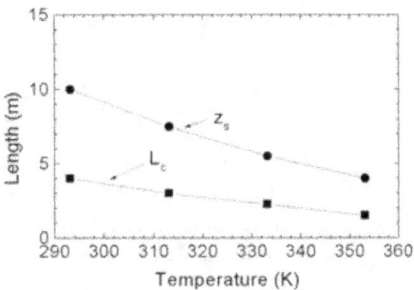

Fig. 6: Variation of the fiber's coupling length L_c for achieving the EMD and length z_s for achieving the SSD with temperature for MH4001 plastic optical fiber.

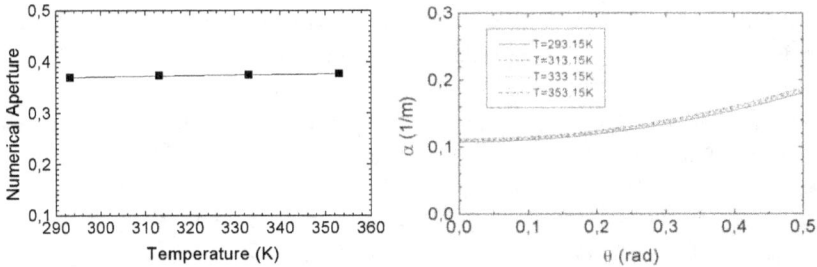

Fig. 7: Variation of the numerical aperture with temperature for MH4001 plastic optical fiber (left). Mode-dependent attenuation ($\alpha'(\theta)$) for MH4001 plastic optical fiber measured at different temperatures (right).

5. Conclusion

Using the power flow equation and experimental measurements, investigated in this article is the state of mode coupling in low-NA (0.3) step-index plastic optical fibers under varied temperature. Results are verified experimentally. It is found that elevated temperatures of low NA SI plastic optical fiber strengthened mode coupling. These properties remained after a year, with the fiber being subjected to environmental temperature variations of more than 35 K. The thermally induced changes of the fiber properties are attributed to the increased intrinsic perturbation effects in the fiber material. Stronger coupling shortens lengths L_c (for achieving the EMD) and z_s (for achieving the SSD). This also results in an increase of the measured bandwidth with increasing fiber temperature. Therefore, bandwidth can be altered by controlling fiber temperature and the results presented in this paper can be used to predict such change.

Acknowledgements

he work described in this paper was supported by the Strategic Research Grant of City University of Hong Kong (Project No. CityU 7004069) and by a grant from Serbian Ministry of Education, Science and Technological Development (Projects No. 171011 and III43008).

References

1. W. Daum, J. Krauser, P. E. Zamzow, O. Ziemann, POF-polymer optical fibers for data communications. Berlin: Springer; 2002.
2. N. Raptis, E. Grivas, E. Pikasis, and D. Syvridis, "Space-time block code based MIMO encoding for large core step index plastic optical fiber transmission systems," Opt. Express 19, 10336–10350 (2011).
3. M. Kovačević and A. Djordjevich, "Variation of modal dispersion and bandwidth with temperature in PMMA based step-index polymer optical fibers," J. Optoelectronics Advan. Mater. 11, 1821–1825(2010).
4. L. Hoffmann, M. S. Müller, S. Krämer, M.Giebel, G.Schwotzer, T. Wieduwilt, "Applications of fibre optic temperature measurement," Proc. Estonian Acad. Sci. Eng. 13, 363–378 (2007).
5. T. Reinsch and J. Henninges, "Temperature dependent characterization of optical fibres for distributed temperature sensing in hot geothermal wells," arXiv:1206.1853v1 [physics.geo-ph] 8 Jun 2012.
6. C.-A. Bunge, R. Kruglov, H. Poisel, "Rayleigh and Mie Scattering in Polymer Optical Fibers," J. of Lightwave Technol., Vol. 24, No. 8, pp. 3137-3146, 2006.
7. A. T. Moraleda, C. V. García, J. Zubia Zaballa and J. Arrue, "A "Temperature Sensor Based on a Polymer Optical Fiber Macro-Bend," Sensors 13, 13076-13089 (2013).

8. Li-Wen Chen, Wei-Hua Lu, Yung-Chuan Chen, "An investigation into power attenuations in deformed polymer optical fibers under high temperature conditions," Opt. Commun. 282,1135–1140(2009).
9. N. Raptis and D. Syvridis, "Bandwidth enhancement of step index plastic optical fibers through a thermal treatment" IEEE Photonics Technol. Lett. 25,1642–1645 (2013).
10. A. F. Garito, J. Wang, and R. Gao, "Effects of random perturbations in plastic optical fibers," Science 281, 962–967 (1998).
11. J. Arrúe, J. Zubía, G. Fuster, and D. Kalymnios, "Light power behaviour when bending plastic optical fibers," in Proc. Inst. Elect. Eng. Optoelectron. 145, 313–318 (1998).
12. M. A. Losada, J. Mateo, I. Garcés, J. Zubía. J. A. Casao, and P. Peréz-Vela, "Analysis of strained plastic optical fibers," IEEE Photon. Technol. Lett. 16, 1513–1515 (2004).
13. S. Savović and A. Djordjevich, "Mode coupling in strained and unstrained step-index plastic optical fibers," Appl. Opt. 45, 6775–6780 (2006).
14. J. Dugas and G. Maurel, "Mode-coupling processes in polymethyl methacrylate-core optical fibers," Appl. Opt. 31, 5069–5079 (1992).
15. A. Djordjevich and S. Savović, "Investigation of mode coupling in step index plastic optical fibers using the power flow equation," IEEE Photon. Technol. Lett. 12, 1489–1491 (2000).
16. M. Eve and J. H. Hannay, "Ray theory and random mode coupling in an optical fibre waveguide, I.," Opt. Quantum Electron. 8, 503–508 (1976).
17. F. Jiménez, J. Arrúe, G. Aldabaldetreku, G. Durana, J. Zubia, O. Ziemann, C.-A. Bunge, "Analysis of a plastic optical fiber-based displacement sensor," OSA Appl. Opt. 46, pp. 6256-6262, 2007.
18. D. Gloge, "Optical power flow in multimode fibers," Bell Syst. Technol. J. 51, 1767–1783 (1972).
19. W. A. Gambling, D. N. Payne, and H. Matsumura, "Mode conversion coefficients in optical fibers," Appl. Opt. 14, 1538–1542 (1975).
20. M. Rousseau and L. Jeunhomme, "Numerical solution of the coupled-power equation in step index optical fibers," IEEE Trans. Microwave Theory Tech. 25, 577–585 (1977).
21. L. Jeunhomme, M. Fraise, and J. P. Pocholle, "Propagation model for long step-index optical fibers," Appl. Opt. 15, 3040–3046 (1976).
22. S. Savović and A. Djordjevich, "Method for calculating the coupling coefficient in step index optical fibers," Appl. Opt.46, 1477–1481 (2007).
23. S. Savović and A. Djordjevich, "Influence of the angle-dependence of mode coupling on optical power distribution in step-index plastic optical fibers," Opt. Laser Technol. 44, 180–184 (2012).
24. S. Savović, M. S. Kovačević, A. Djordjevich, J. S. Bajić, D. Z. Stupar, G. Stepniak, "Mode coupling in low NA plastic optical fibers," Opt. Laser Technol. 60, 85–89 (2014).

Application of the Hankel Transform to Model Misalignment Losses in POFs

M.A. Losada,* J. Mateo A. López

Aragón Institute of Engineering Research, Universidad de Zaragoza, 50018 Zaragoza, Spain.

Corresponding author: alosada@unizar.es

We present a method to assess fiber loss due to misalignments and parameter mismatches in Plastic Optical Fibers that can also be adapted to other multimode fibers. This method is based on the optical power distribution radiated from the end surface of a fiber that can be obtained as the convolution of two circular symmetric functions: the aperture of the fiber and its far field pattern, and whose calculation can be greatly simplified using the Hankel transform. The proposed method is fast, flexible to incorporate combined offsets and its predictions have been successfully compared to experimental measurements. In addition, the same methodology has been adapted to describe not only scalar loss but also its angular dependence.

1. Introduction

Differences or mismatches in the inherent fiber characteristics that occur when tolerances are relaxed during fiber fabrication are an important source of power loss when connecting two fibers. In addition, fiber misalignments are frequent in splices and connectors increasing power loss, independently on the jointing technique. In short-reach networks with multiple connections, slight individual misalignments and mismatches can add together introducing severe global limitations. This effect is particularly accentuated for plastic optical fibers where larger error margins are allowed limiting further the already compromised power budget. Thus, a good model to account for these losses is required to quantify them precisely at the system designing stages [1,2]. So far, most models assess misalignment or mismatches individually using simplifications that render them simple at the cost of accuracy, or require long calculation times to obtain realistic results [3-5]. Here, we propose a method to assess fiber loss due to combined misalignments and parameter mismatches that is based on generally accepted hypotheses [6], but has enough flexibility to incorporate different fiber characteristics. This method is based on the calculation of the optical power distribution radiated from the end surface of a fiber, and its variation as light propagates in free space. First, we showed that this radiated power distribution can be calculated as the convolution of two functions: the aperture of the fiber and its output far field pattern (FFP). As both functions involved are circularly symmetric, the calculation of this 2D convolution can be greatly simplified using the properties of the zero-order Hankel transform [7]. This simplification is a key aspect of our proposal as the use of this transform is critical to reduce complexity and to speed calculations, resulting in a more flexible approach. Once the radiated pattern is obtained it is straightforward to obtain the fraction of power that can be captured by another fiber depending on their relative sizes, apertures and positions. Using this method, we calculated misalignment power losses for combined transversal and longitudinal offsets that were good estimates of experimental results [8]. In addition, a modification of the approach was applied to the upgrade of a matrix connector model to incorporate combined longitudinal and radial misalignments [9]. This connector model is part of the POF matrix propagation framework [10] that has been successfully integrated into commercial software to create a simulation environment that enables the analysis of network designs before their deployment [11].

In this work, we first give a summarized description of our theoretical approach particularized to obtain the power loss for radial and longitudinal misalignments of two identical fibers, but making emphasis on its flexibility to describe other fiber types and to incorporate parameter mismatches and other types of misalignments. Next, we present different applications of the basic approach: first, the surface loss for combined longitudinal and radial offsets is analyzed, demonstrating that their effects are not separable. Then, the high impact of the shape of the FFP over longitudinal misalignment power loss is illustrated using uniform and Gaussian FFPs. Last, we propose a variation of the method to obtain the power transmitted to a shifted receiving fiber as a function of propagation angle where the use of the Hankel transform is paramount to simplify the calculations. These resulting angular dependent functions can be used as a kernel to calculate global power loss. Moreover, as their calculation is independent on the fiber FFP, they provide general information about the effects caused by different misalignment combinations. Finally, the basis and the advantages of this approach are summarized in the Conclusions.

2. Theoretical development

Here, we describe a method to obtain the power loss incurred when there are combined transversal and longitudinal offsets between two identical optical fibers: first, we estimate the radiated power distribution and then, the fraction of the power captured by the receiving fiber.

2.A. Calculation of the radiated power distribution

To calculate the radiated power distribution we assume some previously admitted hypotheses [5,6,8,9]. First, each point in the fiber end surface acts as an independent uncorrelated source whose radiation pattern is given by the fiber FFP. Second, we obtain the total power at any point in the space by adding the scalar contributions from all radiating sources that reach that point. Taking into account the fiber symmetry, we choose to work in cylindrical coordinates

(r, ψ, z) centered on the axis of the emitting fiber, where r is the axial distance, φ is the angle relative to the fiber axis, and z is the distance to the fiber end face along its axis as shown in Fig. 1.

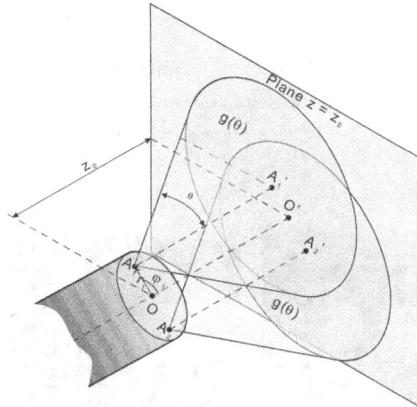

Fig. 1: Schematic of the geometry of power radiation from the fiber end surface. All points (such as A_1, A_2) described in cylindrical coordinates (r, φ, z), emit following the FFP that only depends on the propagation angle θ.

Thus, the total radiated optical power over a point in the plane $z = z_0$ is obtained adding the contributions from all source points such as A_1 and A_2 that is given by the following integral:

$$R(r,z_0) = \int_0^{2\pi} \int_0^{\infty} C(r_1,\varphi_1)g(r_1,\varphi_1,r,\varphi,z_0)r_1\,dr_1\,d\varphi_1 \qquad (1)$$

Using the fact that the FFP only depends on the propagation angle, θ, and that $C(r_1, \varphi_1)$ is a circle of radius a and unit amplitude, we obtain the following integral, using basic trigonometry relationships:

$$R(r,z_0) = \int_0^{2\pi} \int_0^{a} g\left(\tan^{-1}\left[\frac{\sqrt{r_1^2 + r^2 - 2r_1 r \cos(\varphi_1 - \varphi)}}{z_0}\right]\right) r_1\,dr_1\,d\varphi_1 \qquad (2)$$

Notice that even though all functions have rotational symmetry, there is still a dependence on the angles φ and φ_1 that makes this integral tedious to calculate. However, Eq. (2) is exactly the two-dimensional convolution of the circle and the FFP expressed in cylindrical coordinates [7]. As both functions are circular symmetric, the calculation can be simplified using the Hankel transform of order 0 as:

$$R(r,z_0) = C(\vec{r}) **g(\vec{r}) = C(r) **g\left(\frac{r}{z_0}\right) = z_0^2 \cdot H^{-1}\left\{\frac{aJ_1(2\pi a\rho) \cdot G(z_0\rho)}{\rho}\right\}, \qquad (3)$$

where H^{-1} is the inverse zero-order Hankel transform, $G(\rho)$ is the transform of $g(r)$, and $\frac{a \cdot J_1(2\pi a \rho)}{\rho}$ is the transform of a uniform distribution. Although the advantages of this approach may not be immediately evident, we have to consider that most functions proposed to model the radial profile of the FFP (uniform, Gaussian, etc.) have tabulated analytical transforms. Thus, in most cases, only the inverse transform remains to be obtained using one of the many fast algorithms implemented in most commercial software packages. In addition, our first hypothesis that considers that all points on the fiber end surface radiate the same amount of light can be relaxed to accommodate graded-index fibers using their index profiles instead of the circular aperture [5].

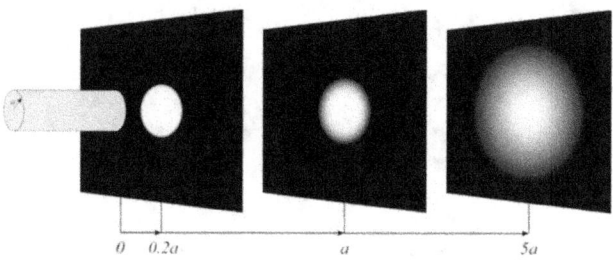

Fig. 2: Radiated pattern at different distances from the fiber end calculated using a Gaussian FFP.

In Fig. 2, image representations of the radiated patterns calculated using a Gaussian FFP are shown at different distances from the fiber. Near the fiber end ($z < a$) the radiation pattern resembles the circular fiber aperture, both in size and in shape. As the distance from the radiating fiber increases power spreads radially, and the pattern changes gradually to assume the shape of the far field pattern. At distances longer than 50a (not shown) the simulated pattern coincides with the fiber FFP. We will illustrate later how the shape of the FFP used to calculate the radiation pattern determines not only its profile but also its dependence with length, both of which have an impact on the predicted misalignment losses. Also, we want to stress here the method flexibility as it accommodates not only analytical FFPs, but can also incorporate experimental or non-analytical FFPs provided they have circular symmetry and can be described by their radial profiles.

2.B. *Estimation of the alignment power loss*

Once the radiated power distribution $R(r, z_0)$ has been obtained for a particular case using Eq. 3, we can obtain the misalignment loss assuming that the power captured by the receiving fiber is the fraction of power radiated from the first fiber that overlaps with the core surface of the second fiber as shown in Fig. 3. The power coupled is calculated integrating the radiated pattern over the surface of the second fiber C':

$$P(r_0, z_0) = \iint_{C'} R(r, z_0) r \, dr \, d\varphi \tag{4}$$

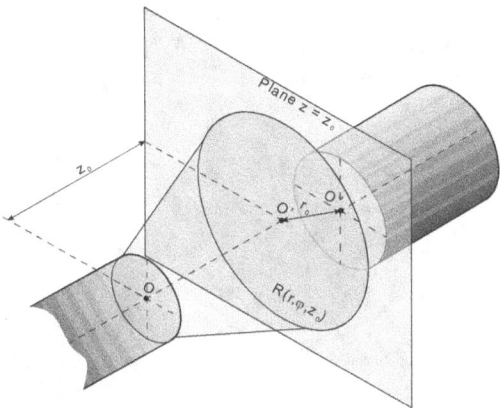

Fig. 3: Schematic that illustrates radial and longitudinal offsets between two fibers of different radii.

Notice that the second fiber can have different size (circle of radius b) or even different shape (ellipsoidal) than the radiating fiber as the restriction of circular symmetry is no longer required. In fact, our methodology can also be applied to estimate the light that is coupled into other devices with different shapes such as a detector shifted from the fiber end.

We introduce another function to account for the difference in the power captured by the receiving fiber, depending on its angle of incidence. We call it differential efficiency function and model it using the FFP of the receiving fiber. Therefore, its effects are introduced in the calculation by simply changing $g(\theta)$ that now will be the product of the FFP of the emitting fiber, $g_1(\theta)$, and the FFP of the receiving fiber, $g_2(\theta)$: $g(\theta) = g_1(\theta) \cdot g_2(\theta)$. In this way, the loss of joining fibers with different apertures can be obtained in combination with misalignments. Otherwise, if both fibers are equal, $g(\theta)$ will be the squared fiber FFP. In any case, this enhancement does not introduce further calculations due to the use of the Hankel transform.

3. Results

3.A. Combined radial and longitudinal misalignment power loss

To demonstrate how the proposed method is able to predict losses for combined misalignments, Fig. 4 shows the results for longitudinal and radial offsets of fibers with the same properties (radius, shape, NA, etc.) calculated using a Gaussian FFP. The middle graph shows the contour plot of power losses in dB and illustrates the higher impact of radial shifts over global loss. On the left, the variation of loss with longitudinal misalignment is shown for several radial offsets. On the right, loss as a function of radial misalignment is represented for the fibers at different longitudinal separations. These curves show how the effects of these two misalignments are not independent which makes it necessary to assess their combined effects. In general, introducing radial shifts to the longitudinal separation always increases power loss, but the power spatial spread for longer fiber distances produces decreases losses at the extreme radial shifts.

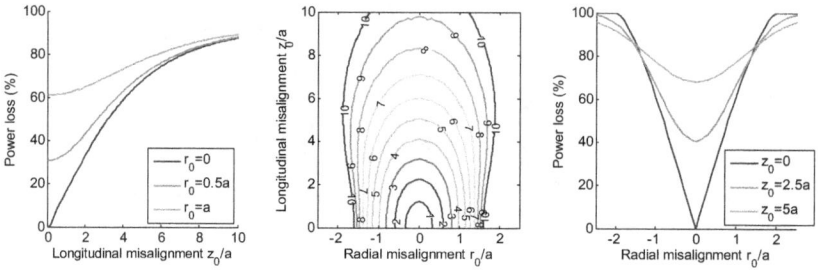

Fig. 4: Losses for combined radial and longitudinal misalignments: percentage of power loss versus longitudinal separation for three radial offsets (left); contour plot of combined misalignment loss in dB (center); power loss percentage versus radial misalignment for three longitudinal shifts (right).

3.B. Influence of the FFP on misalignment losses

In Fig. 5, the radiated power distribution is shown at different distances from the fiber calculated for two different FFPs. On the left, power distributions are shown for a uniform FFP and in the middle, for a Gaussian FFP. Notice that the assumption of a uniform FFP is not equivalent to consider a uniform radiated power distribution at any distance z. The curves show that, for distances above the fiber radius, the power spreads very differently depending on the FFP. These differences translate to the dependence of loss versus longitudinal offset of two identical fibers shown on the right for the two cases. The uniform distribution predicts a steeper increase of power loss with fiber distance than the more realistic Gaussian distribution. These predictions are consistent with those found when comparing experimental data with the prevailing model based on a uniform power distribution spread at the rate given by the fiber NA [3; pp. 260,8] although the later predicts an even steeper increase of power loss. Thus, we have shown how the shape of the FFP used to calculate the radiation pattern determines not only its profile but also its dependence with length, both of which have an impact on the predicted misalignment losses. This is particularly important for POFs whose power distribution is strongly dependent on launching conditions and fiber type, what adds to their strong mode coupling and differential attenuation that produce rapid changes in power distribution with propagation distance [12]. In addition, it has been demonstrated that power distribution can be easily altered by curvatures and other localized disturbances [13]. Therefore, the effects of misalignments in POFs can change drastically for different conditions and also depend on their position in the line [9]. As any change in the shape of the FFP will imply to re-calculate the radiation power distribution using Eq. (3) and then, applying Eq. (4) to obtain actual losses, in the next subsection we describe a framework that does not require advanced knowledge of the shape of the FFP.

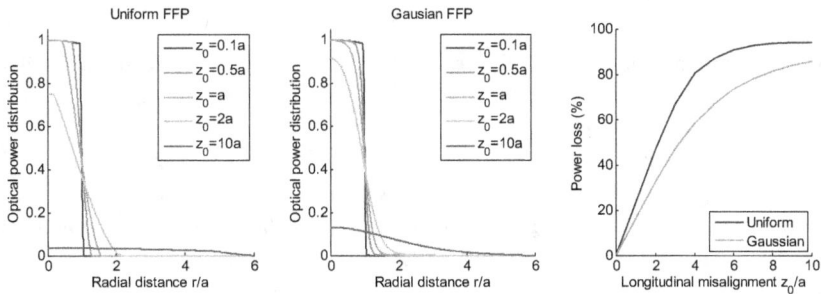

Fig. 5: Influence of the FFP on longitudinal losses: radiated profiles at different distances from the fiber end for uniform FFP (left), and Gaussian FFP (center); loss percentage versus longitudinal misalignment for both FFPs (right).

3.C. Angular-dependent power loss

Here, our aim is to offer a new tool to describe the changes in power distribution that are caused by combined longitudinal and transversal shifts for identical fibers, although it could be generalized also to account for parameter mismatches and other misalignments. Thus, we propose an approach similar to that described before, keeping the same assumptions but where it is not necessary to know either the emitting or the receiving fiber FFPs. In this case, we calculate the proportion of light that, exiting the radiating fiber only at a given angle θ_r, is able to reach the receiving shifted fiber. This is equivalent to assume that each point of the fiber radiates light in a very narrow angular range centered at θ_r, whose projection onto a perpendicular plane is a very narrow ring instead of a circle. The calculation flow is similar to that described before but at the end we obtain a function $P(\theta_r, r_0, z_0)$ that represents the proportion of power propagating at angle θ_r that is transferred to the receiving fiber when it is shifted at (r_0, z_0). Therefore, this function gives, not only the absolute power loss produced by fiber misalignments, but also its angular variation independently from the FFP, and thus, allows to draw general conclusions regarding the effect of different shift combinations. Fig. 6 shows this function for different longitudinal offsets combined to radial misalignments.

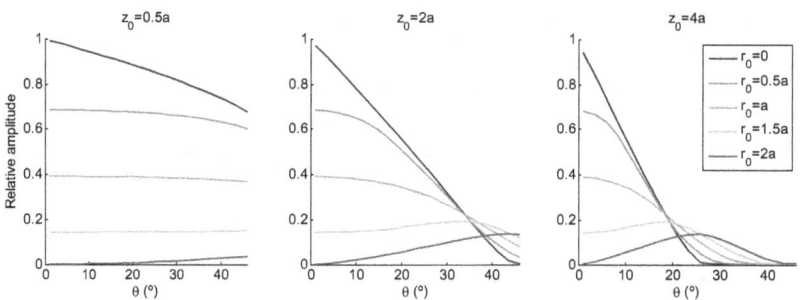

Fig. 6: Function $P(\theta_r, r_0, z_0)$ for different combinations of longitudinal and radial shifts.

The graphs show how, when there is no radial shift, power at higher angles is reduced when increasing z distances. This finding is consistent with previous reports that have indicated the effect of longitudinal shifts acting as a spatial filter that is able to increase bandwidth when placed near detection [9]. For small z offsets and r shifts above $0.5a$, losses are greater but relatively constant with the propagation angle. However, if the z shift increases, the angular dependence is no longer constant and the relative power coupled from lower and higher angles changes depends on the transversal offset. In fact, for a radial shifts larger than the fiber radii power from higher angles is more effectively coupled to the other fiber than power transmitted near the axis ($0°$). This is reasonable as, when the centers of the fibers are separated by large transversal shifts, only light radiated at the highest angles is able to reach the receiving fiber. As it was measured that misalignments do not increase diffusion but only introduce this angular dependent power loss, these functions have been applied to upgrade a matrix connector model generalized to incorporate radial and longitudinal misalignments [8].

In addition, $P(\theta_r, r_0, z_0)$ can be treated as a kernel to obtain the power transferred to the receiving fiber for a given misalignment pair (r_0, z_0) using Eq. (4). Now, once the user knows how to implement $g(\theta)$, the only calculation required to obtain the power coupled into a fiber placed at (r_0, z_0) will be the following integral:

$$P(r_0, z_0) = \int_0^{\pi/2} g(\theta) P(\theta, r_0, z_0) \sin(\theta) \, d\theta \tag{5}$$

The pre-calculation of $P(\theta_r, r_0, z_0)$ for usual offset values to obtain power losses can help when calculation time is critical such as in real time simulations.

4. Conclusion

We propose that the evolution of the radiated power distribution as light propagates in free space can be calculated as a 2D convolution simplified using the Hankel transform. The radiated pattern can be thus simply obtained for different fiber types and conditions at different distances from the fiber. Using this power distribution the fraction of power that can be captured by another fiber depending on their relative positions can be obtained to predict misalignment losses. Our method is faster than others and more flexible than most, as parameters mismatches such as radius and numerical aperture differences can be easily incorporated without introducing further complexity in the calculations. A variation of the method permits to obtain the angular-dependent power coupling that can be used to assess the effects of different misalignment combinations over the distribution of the coupled power without assuming the shape of the FFP of the fibers involved. This paradigm has already been applied to upgrade a connector matrix obtained from experimental data and included in a propagation model that is integrated into simulation software.

Acknowledgements

This work was supported by the Spanish Ministerio de Economía y Competitividad under project TEC201237983C0303.

References

1. U. H. P. Fischer, M. Haupt, and M. Joncic, 'Optical Transmission Systems Using Polymeric Fibers', in Optoelectronics Devices and Applications, P. Predeep, ed., 445-468 (InTech, 2011).
2. D. H. Richards, M. A. Losada, N. Antoniades, A. López, J. Mateo, X. Jiang and N. Madamopoulos, 'Modeling Methodology for Engineering SI-POF and Connectors in an Avionics System', J. Lightw. Technol., 31(3), 468-475 (2013).
3. O. Ziemann, J. Krauser, P. E. Zamzow and W. Daum, POF Handbook, 2nd ed. (Springer, 2008).
4. G. Aldabaldetreku, G. Durana, J. Zubia, J. Arrue, H. Poisel and M. A. Losada, 'Investigation and comparison of analytical, numerical and experimentally measured coupling losses for multi-step index optical fibers', Opt. Express, 13(11), 4012-4036 (2005).
5. J. Xu, M. Bloos, and H. Poisel, 'Improved modelling of connector losses for SI-POF based on exact values for the radiance at fiber end faces', in Proc. Int. Conf. on Transparent Optical Networks, Graz, Austria (2014).
6. S. Werzinger, C. A. Bunge, S. Loquai, and O. Ziemann, 'An analytic connector loss model for step index polymer optical fiber links', J. Lightw. Technol., 31(16), 27692776 (2013).
7. N. Badour, 'Operational and convolution properties of two-dimensional Fourier transforms in polar coordinates', J. Opt Soc. Am. A, 26(8), 1767-1777 (2009).
8. J. Mateo, M. A. Losada, N. Antoniades, D. Richards, A. López, and J. Zubia, 'Connector misalignment matrix model', in Proc. Intl. Conf. on Plastic Optical Fibres and Applications, Atlanta, USA (2012).
9. J. Mateo, M.A. Losada and A. López, 'POF misalignment model based on the calculation of the radiation pattern using the Hankel transform', Opt. Express, 23(6), 80618072 (2015).

10. J. Mateo, M.A. Losada, J. Zubia, 'Frequency response in step index plastic optical fibers obtained from the generalized power flow equation', Opt. Express, 17(4), 2850-2860 (2009).
11. A. Alcoceba, A. López, M. A. Losada, J. Mateo and C. Vázquez, 'Building a simulation framework for POF data links', in Proc. Intl. Conf. on Plastic Optical Fibres and Applications, Nuremberg, Germany (2015).
12. J. Mateo, M. A. Losada, and I. Garcés, 'Global characterization of optical power propagation in step index plastic optical fibers', Opt. Express, 14(20), 9028-9035 (2006).
13. M. A. Losada, J. Mateo, and J.J. Martínez-Muro, 'Assessment of the impact of localized disturbances on SI-POF transmission using a matrix propagation model', IOP Journal of Optics, 13, 055406-055412 (2011).

Time-Domain Solution to the Power-Flow Equation and Possible Applications

M. Gehrke,[1,*] S. Loquai,[1] O. Ziemann,[1] B. Schmauss[2]

[1]*Polymer Optical Fiber Application Center,*
Technische Hochschule Nürnberg, 90489 Nürnberg, Germany.

[2]*Institute of Microwaves and Photonics (LHFT),*
Friedrich-Alexander-Universität Erlangen-Nürnberg, 91054 Erlangen, Germany.

*Corresponding author: martin.gehrke@pofac.th-nuernberg.de

This paper presents a simulation approach to obtain time domain backscatter signals of large core step index polymer optical fibers (SI-POF) based on the time dependent power flow equation. In this model three major fiber impairments mode coupling, modal attenuation and modal dispersion are taken into account as functions of the axial propagation angle. Other aspects such as chromatic and material dispersion are neglected due to less significant consequences regarding the transmission characteristics of the fiber.

1. Introduction

The characteristics of multimode fibers, like transfer function and attenuation, have been widely analyzed in both theoretical and experimental manner. This work is mainly based upon previous research by Losada et al. [1], who devised the matrix implementation of Gloge's original description of the power flow equation [2] to calculate frequency and impulse responses of POF. Breyer et al. [3], Drljaca et al. [4] and Gloge [5] have previously published papers about numerical and analytical solutions of the power flow Eq. as well.

Here we expand this solution with a model, which can additionally predict the angular power distribution of guided reflected light based upon the analysis of backward propagating signals. These are primarily based upon Rayleigh scattering, which occurs during interaction between propagating light and fiber inhomogeneities such as density fluctuations inherent to amorphous material like PMMA. This simulation aspect can be used to predict optical time domain reflectometry (OTDR) measurements such as attenuation over fiber length.

2. Model description

In this chapter the already known numerical solution of the power flow Eq. is quickly recapitulated. Afterwards the problem regarding noise generation due to the leakage effect of the Fourier Transformation is explained. To circumvent this effect a solution of the time dependent power flow Eq. directly in the time domain is derived. Additionally the implementation of the backscatter simulation aspect and its underlying theory is explained.

This work often refers to the well known term mode to describe the distribution of the electromagnetic field induced by propagating light in a multimode fiber. Hereby the three dimensional propagation angle of a mode, which is often commonly regarded as a light ray, is given by its propagation direction and a line parallel to the fiber axis. It will be called θ in the following document. A graphical depiction of the angle in question can be found in Snyder [6].

2.A. Frequency domain based solution of the power flow equation

The power flow Eq. was originally contrived by Gloge [5] to describe and investigate the behavior of the electromagnetic field in optical fibers without solving the classical wave equations which is inherently difficult and requires huge effort and extensive processing power. This has still not been performed today, 40 years later. To circumvent these restraints, Gloge [2] used an angular dependent attenuation and diffusion process, to describe modal attenuation and mode mixing behavior of multimode fibers. This gives the basic form of the power flow equation:

$$dP = -\alpha(\theta)P dz + \frac{1}{\theta} \cdot \frac{\partial}{\partial \theta} \cdot \left(\theta \cdot D(\theta) \cdot \frac{\partial P}{\partial \theta} \right) dz \tag{1}$$

The modal attenuation is described by $\alpha(\theta)$ and the mode coupling, which is limited to neighboring modes, is given by $D(\theta)$. A temporal dependency can be introduced, if P is a function of distance (z) and time (t), by using the total differential

$$dP = \frac{\partial P}{\partial z} dz + \frac{\partial P}{\partial t} dt. \tag{2}$$

A temporal dependency can be introduced by using the mode delay relation, which is incurred by the different zigzag travel paths modes are subjected to inside a fiber depending on their propagation angle. The relation of the path lengths is given by the simple equation:

$$z(\theta) = \frac{z(0°)}{\cos(\theta)} \tag{3}$$

This leads to the group velocity $v(\theta)$ of a mode with the propagation angle θ given by the differential

$$v(\theta) = \frac{dz}{dt} = \frac{c}{n \cdot \cos(\theta)}, \tag{4}$$

where c is the speed of light and n is the refractive index.

Substituting Eq. (2) and (4) into (1) results in the basic form of the time dependent power flow equation:

$$\frac{\partial P(\theta,z,t)}{\partial z} = -\alpha(\theta)P(\theta,z,t) - \frac{n}{c\cdot\cos(\theta)}\cdot\frac{\partial P(\theta,z,t)}{\partial t}$$
$$+ \frac{1}{\theta}\cdot\frac{\partial}{\partial\theta}\cdot\left(\theta\cdot D(\theta)\cdot\frac{\partial P(\theta,z,t)}{\partial\theta}\right) \quad (5)$$

Using the Fourier transformation

$$p(\theta,z,\omega) = \mathfrak{F}\left\{P(\theta,z,t)\right\} \quad (6)$$

the time differential can be eliminated and is exchanged with a frequency dependency. This leads to

$$\frac{\partial p(\theta,z,\omega)}{\partial z} = -\left(\alpha(\theta) + \frac{n}{c\cdot\cos(\theta)}\cdot j\omega\right)p(\theta,z,\omega)$$
$$+ \frac{1}{\theta}\cdot\frac{\partial}{\partial\theta}\cdot\left(\theta\cdot D(\theta)\cdot\frac{\partial p(\theta,z,\omega)}{\partial\theta}\right), \quad (7)$$

which can be solved analytically [5] or numerically [1,3,4]. These publications can be consulted for further reading.

2.B. Leakage effect

The solutions based on Eq. (7) require first a transformation from the time into the frequency domain, before the actual propagation algorithm is applied. Hereby the different angular dependent mode delays are applied using a phase shift in the frequency domain, since the phase in the frequency domain correlates with the time in the time domain according to the well known shift theorem [7]:

$$\mathfrak{F}\left\{f(t-t_0)\right\} = \mathfrak{F}\left\{f(t)\right\}\cdot e^{-j\omega t_0} \quad (8)$$

During the simulation this phase shift is only limited by the precision of the used data type for the calculation. Upon re-transformation of the complex signal using the inverse Fourier transformation to again yield a time domain signal the data points have to be sorted into their respective time bin. Since the time delay is caused by different mode zigzag travel paths, which are described by a nonlinear function $(\cos(\theta))$, at least for some angles a discretization problem will be incurred. This is commonly called a leakage effect.

This problem is depicted in Fig. 1. In this graph one can see a Dirac impulse and the corresponding phase of the Fourier transform. By changing the slope of the phase the inverse transformed signal can be moved in the time domain. If the factor by which it is multiplied is correctly chosen (i.e. multiples of $\frac{f_s}{N}$, where f_s is the sampling frequency and N is the length of the data) then the corresponding time domain signal fits into the time bins. This can be seen in the second row of Fig. 1. If however a fractured multiple of $\frac{f_s}{N}$ is chosen to modify the

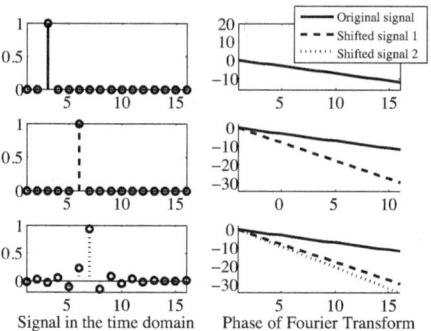

Fig. 1: Three shifted time domain signals and their corresponding Fourier transform phase

phase, then the corresponding signal doesn't fit into the appropriate time bins, as is shown in the bottom row. The discretization will then lead to the mentioned noise problematic.

Zero padding can mitigate this effect to some extend, however there will always be modes where the effect is visible. This creates a noise floor in the simulated power distribution data, which is smaller by about 10^5 than the actual signal. Usually one might ignore these inaccuracies, however since the backscattered data, which will be simulated, is also significantly below the forward propagating power level, the noise floor influences the result, which cannot be tolerated.

2.C. Time domain based solution of the power flow equation

Instead of solving in the frequency domain to eliminate the temporal differential, the power flow equation (5) could also be solved directly in the time domain. Upon inspection of (5) it can be seen, that the only part of the Eq. that actually bears a time dependency is the mode delay. Neither the modal attenuation nor the modal diffusion are affected by the inclusion of the time domain. If the temporal scale is assumed to be a collection of single point angular power distributions (i.e. time-independent) the result is an accumulation of distributions as described by equation (1). In order to correctly account for the mode delay the complete space-time signal has to be reconstructed and the modal delay has to be applied after each incrementation step.

The result is a two step process, where the stationary power distribution (i.e. for one time-point of the input signal) for every distance step Δz has to be calculated using the appropriate attenuation and diffusion process and afterwards corrected with the proper time delay for every mode. Using the reasonable assumption of time independence regarding the propagation processes for one small distance step, we obtain the power propagation differential equation

$$\frac{\partial P(\theta,z,t)}{\partial z} = -\alpha(\theta)P(\theta,z,t)$$
$$+ \frac{1}{\theta} \cdot \frac{\partial}{\partial \theta} \cdot \left(\theta \cdot D(\theta) \cdot \frac{\partial P(\theta,z,t)}{\partial \theta}\right), \qquad (9)$$

which can bo solved by applying an explicit finite-difference scheme to the differentials. We used a forward difference for the spatial derivative and central differences for the angular derivatives. This yields the numerically solvable equation

$$P_{it}(\theta, z+\Delta z, t) = \left(1 - \alpha(\theta) - \frac{2D(\theta)}{\Delta\theta^2}\right)\Delta z \cdot P(\theta,z,t)$$
$$+ \left(\frac{D(\theta)}{\Delta\theta} - \frac{1}{2}\cdot\frac{D(\theta)}{\theta} - \frac{1}{4}\cdot\frac{D(\theta+\Delta\theta)-D(\theta-\Delta\theta)}{\Delta\theta}\right)$$
$$\cdot \frac{\Delta z}{\Delta\theta} \cdot P(\theta-\Delta\theta,z,t)$$
$$+ \left(\frac{D(\theta)}{\Delta\theta} + \frac{1}{2}\cdot\frac{D(\theta)}{\theta} + \frac{1}{4}\cdot\frac{D(\theta+\Delta\theta)-D(\theta-\Delta\theta)}{\Delta\theta}\right)$$
$$\cdot \frac{\Delta z}{\Delta\theta} \cdot P(\theta+\Delta\theta,z,t), \qquad (10)$$

with the boundary conditions being defined in a way without a possible power loss due to diffusion, since this process is defined to be inherently lossless. As previously stated the mode delay has to be applied immediately after each distance step as soon as the power distribution $P_{it}(\theta,z+\Delta z,t)$ has been calculated. Otherwise the assumption of temporal independence, which was used to obtain Eq. (9) does not hold true. The correct values can be obtained using the relative distance difference between modes

$$z_{diff}(\theta) = \Delta z \cdot \left(\frac{1}{\cos(\theta)} - 1\right), \qquad (11)$$

which subsequently yields the relative mode delay

$$t_{delay}(\theta) = \frac{n \cdot \Delta z}{c} \cdot \left(\frac{1}{\cos(\theta)} - 1\right). \qquad (12)$$

This two step process of first propagating a space time power distribution one distance step and then applying the correct mode delay can be expressed as well. The first of those operations can mathematically be described as a matrix multiplication

$$\underline{P}_{it} = \underline{T} * \underline{P}, \qquad (13)$$

where the elements of the tri diagonal Matrix $\underline{T}$ can be directly obtained from Eq. (10). While the resulting Matrix $\underline{P}_{it}$ is now correctly modified for one incremental distance step regarding attenuation and diffusion, it still requires a conversion to implement the mode delay. This can easily be achieved by shifting each row (i.e. modified the time position of each angle increment) by the correct amount:

$$P(k \cdot \Delta\theta, z + \Delta z, t) = \text{shift}\left(P_{it}\left(k \cdot \Delta\theta, z + \Delta z, t\right)\right),$$
$$\text{with } k = 1, ..., k_{max} \tag{14}$$

This concludes one iteration step. To simulate any desired fiber length the two step process can easily be repeated any number of times.

The required relative mode delay for the shifting process can be calculated using Eq. (12). As such the time steps depend on the defined spatial and angular resolutions (Δz and $\Delta\theta$ respectively). Due to the discretization a maximum error of $\pm 0.5\Delta t$ is incurred, which is shown in Fig. 2 for $\Delta\theta = 0.02°, \Delta z = 1\,\text{mm}$ and $\Delta t = t_{delay}(\Delta\theta) = 1.1624 \times 10^{-16}\,\text{s}$.[3] Essentially this exchanges the leakage effect with a rounding error. This ensures no noise generation by the algorithm, which is preferable.

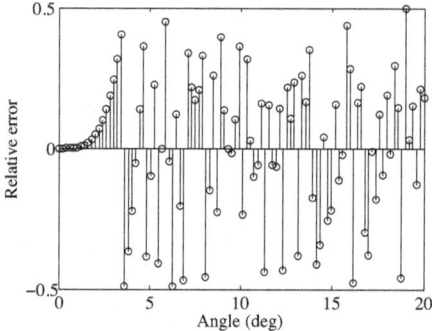

Fig. 2: Relative error caused by by the process of discretization over propagation angle

2.D. Backscatter theory

Commercially available POFs typically show a total transmission loss of $150\,\text{dB}\,\text{km}^{-1}$ at a wavelength of 650 nm [8]. This attenuation is partly caused by intrinsic absorption losses, which are molecular vibrations or electronic transitions, or by extrinsic losses such as impurities caused by contaminants or variations of the core diameter.

A part of this total loss is caused by scattering. As much as 95 % of the total scattering losses of the core of the fiber are caused by Rayleigh scattering, which makes this the main aspect to consider [9]. It is caused by sub-wavelength density fluctuations, which result in a slightly irregular composition of the transport medium. This form of scattering is an intrinsic process inherent to amorphous materials like polymers [10].

---

[3]choosing $\Delta t = t_{delay}(\Delta\theta)$ ensures that the lowest order mode can be discerned from the base mode

The oscillating electrical field of the incident electromagnetic wave used for transmission will inevitably interact with any molecules encountered during its travel path. Especially the electrons of the long chain molecules within the POF are displaced constantly and will oscillate at the same frequency as the incident wave, which means they can be considered electrical dipoles. In dense media like POF we regard them as an isotropic dielectric sphere, which scatters light in all directions non coherently. [11]

In order to quantify light scattering due to density fluctuations one can use the turbidity formula [12], which is given by equation (15) for an amorphous material, which keeps its liquid molecular structure from the melting process during manufacturing, like POF:

$$\tau = C \cdot \frac{8\pi}{3\lambda^4} \cdot \left(\frac{\left(n^2 - 1\right)\left(n^2 - 2\right)}{3} \right) \cdot kT_{gt}\beta_T \tag{15}$$

A strong dependency on the wavelength of the incident light can be seen by the inverse proportionality to the forth power of λ. T_{gt} is the glass transition temperature (378 K [13]), which expresses the temperature at which the liquid molecules are fixed inside the material, k the Boltzmann constant, n the refractive index of the core of the fiber, β_T the isothermal compressibility (3.6×10^{-10} m^2/N [14]) and C the Canbannes factor (1.1), which corrects for the molecular anisotropy of the polymers contained inside the fiber.

The fraction of recaptured power for a step index fiber [15] is given by

$$S \cong \frac{\pi NA^2}{4\pi n^2} = \frac{(NA)^2}{4n^2}, \tag{16}$$

which assumes the Rayleigh scattering to be isotropic, to describe the angle of accepted modes. ($\frac{NA}{n}$) constitutes the half angle of the cone of captured rays). This has been considered in the simulation by using an isotropic scatter distribution.

2.E. Backscatter simulation implementation

Using a power distribution, which can be simulated with regards to angular and temporal distributions with the solution to the time dependent power flow Eq. as shown in section 2.C, at a specific fiber length the level (based on equation (15)) and shape (isotropic) of backscattered power caused by Rayleigh scattering can be calculated. This in turn essentially returns another power distribution over time and angle P_{bs}, which can be used as the new input parameter (in essence a fictive light source) for another propagation simulation, again using the time dependent power flow equation. This time however the propagation is in backwards direction (i.e. towards the original beginning of the fiber). Since the signal traverses the same fiber as before, it follows, that it is subjugated to the same diffusion and attenuation processes in the same way as they would apply to forward propagating signals.

This allows looking at the shape of one reflected distribution at the fiber input $\underline{P}_{fi}$. Subsequently however if a pulse is simulated for each incremental distance step and its corresponding reflection signal is additionally generated and propagated backwards, a superposition of all the distribution shapes, while correctly accounting for their respective time delays, can be achieved using the equation:

$$P_{total}(\theta, t) = \sum_{i=1}^{z_{max}} P_{fi}(\theta, i \cdot \Delta z, t) \tag{17}$$

The result shows the complete backscattered power flow at the beginning of the fiber over time and angle. If the time scale is converted to distance (the delay correlates with the originating distance of the reflected pulse) we obtain what is essentially an OTDR trace just with additional angular resolution. The classic power level over distance trace can also be obtained by integrating over the angle.

3. Simulation results

In this chapter several simulation results using the aforementioned model are shown and compared with previously published results to show the validity of this approach. A backward simulation result will be presented as well to show the general concept.

3.A. Forward propagation simulation

With the previously mentioned numerical solutions to the power flow Eq. it is possible to simulate the power distribution along the length of a fiber using a starting distribution form a light emitting source. [1, 3, 4] have already reported on their simulation results, which are used as a reference point to compare our newly acquired simulations to. The most recent coefficients for the diffusion and attenuation functions reported by Mundus et al. [16] are used. Fig. 3 shows on the left the results obtained using the time domain based numerical

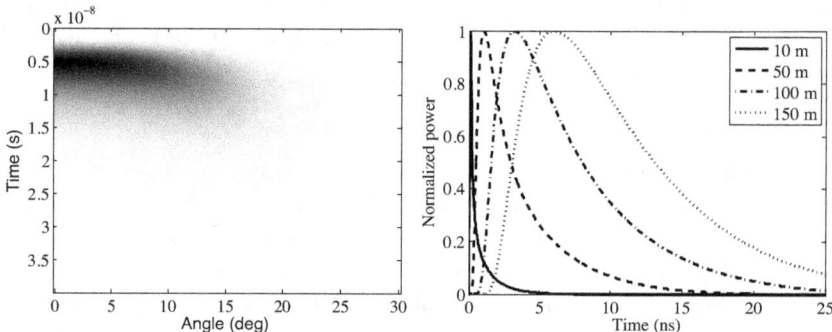

Fig. 3: Space time power distribution of 150 m PGU fiber (left) and impulse responses for four lengths of Eska GH fiber (right)

solution of the time dependent power flow equation. Fig. 3 depicts on the right the space time power distribution, which arises for a Toray PGU-FB1000 fiber excited with a Dirac impulse

($\Delta t = 1 \times 10^{-12}$ s) in the time domain and a Gaussian angular distribution with a numerical aperture of 0.2 according to equation (18). The grayscale represents the normalized power, with black representing the highest and white the lowest values. Losada et al. [1] obtained results for the same distance as well.

$$P(\theta, z = 0, t = 0) = \exp\left(-\ln 2 \cdot \left(\frac{\theta - \theta_0}{\arcsin(NA)}\right)\right) \qquad (18)$$

Fig. 3 (right) shows four impulse responses for different lengths (10 m, 50 m, 100 m and 150 m) of Mitsubishi Eska-Premier GH40017 fiber. The same four impulse responses have also been calculated by Breyer [3], who used a Crank-Nicolson scheme to solve the power flow Eq. numerically in the frequency domain ($\Delta t = 20 \times 10^{-12}$ s and numerical aperture 0.17).

3.B. Backscattered powerflow simulation

As previously stated in section 2.E the simulation process can also be used to create OTDR-like traces of power level differences over distance. The results can be obtained with an additional angular resolution, if so desired, as well.

Fig. 4 shows two of these backscatter simulation results of a Mitsubishi Eska-Premier GH40017 fiber obtained with the discussed simulation method. The left graph of Fig. 4) shows the reflected power over fiber distance, whereas on the right-hand side the backscattered power distribution is depicted over the length of the fiber with an additional angular resolution.

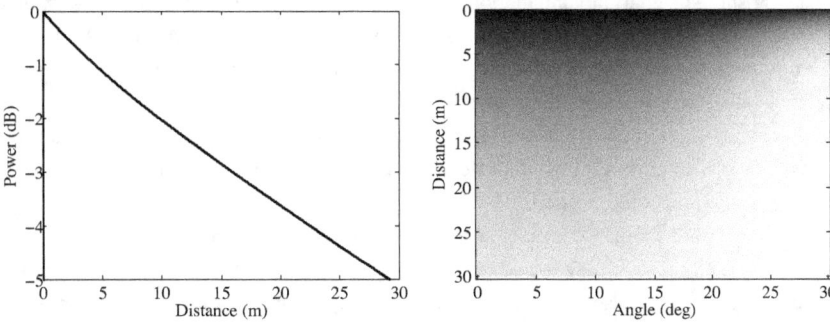

Fig. 4: Backscattered power simulations over distance (left) and angle (right) for Eska GH fiber

4. Discussion

Figure 3 shows already known results from previously published works re-simulated with the new approach based in the time domain. Upon comparison the results show a very similar result. Fig. 3 (left), whose equivalent can be found in [1], shows virtually no difference at all.

A very similar pattern can be observed as well with Fig. 3 on the right. The behavior of the right slope of the impulse response shows a slightly steeper drop off in Breyer's findings [3]. The maximum for 150 m is also slightly shifted to the right (i.e. more delay) in our simulations. The reason for these minute differences can be attributed to the marginally different diffusion and attenuation coefficients. This slight deviation notwithstanding the time domain based approach to solve the time dependent power flow Eq. delivers the same results as previous solutions in the frequency domain [1, 3, 4].

The main difference between the algorithms is of course the usage of different domains for the solution. Strictly talking about the forward direction the frequency based algorithms show a better performance since most of the processing time is expended on the calculation of the transfer matrix. The propagation by matrix multiplication is performed in the frequency domain, which means two Fourier transformations are also required to convert to and back from the frequency domain. Our algorithm requires a calculation of the power distribution of every distance increment step and thus creates a lot of useful interim results, which takes more time.

Another advantage of this new approach is its noise avoidance in the time domain. The mode delay in previous implementations is applied in the frequency domain as a phase shift of the signal. The precision of the phase change is only limited by the data type used in the calculation. This minimal phase shift however inevitably introduces a leakage effect upon retransformation since the constant time bins are spaced more widely and do not match the only slightly shifted signal. In combination with the diffusion process a noise floor is created beneath the actual signal. Our algorithm uses a straightforward shift, which can only shift to the next time increment, which inevitably will cause a rounding imprecision. This however successfully eliminates the noise floor problematic from the process.

Lastly it should be noted, that due to human nature processes in the time domain are significantly easier to grasp than similar calculations in the frequency domain, which is another advantage of the proposed approach.

The first backward simulation results in section 3.B show a promising behavior as well. The overall attenuation shown on the left of Fig. 4 is $160\,\text{dB}\,\text{km}^{-1}$, which seems reasonable for the used fiber. Fig. 4 (right) also shows a reasonably quickly diminishing impact of higher order modes, which certainly should be expected due to the higher attenuation for those propagation angles. However no angular dependent measurements have been performed yet and thus the results cannot be properly validated as of now.

5. Conclusion

We have proposed a new solution to the power flow equation, which is directly executed in the time domain instead of transforming to the frequency domain and back. This results in a fast, robust and easy to understand way of obtaining space-time power distributions in optical fibers. The simulation results have been verified with comparisons to previously published data and the performance for multiple forward simulation distances is comparable as well. In addition to the power distributions the algorithm additionally allows the calculation of other fiber parameters such as bandwidth and pulse spreading over fiber length. The algorithm is inherently well suited for backscatter simulation applications as well. Backscatter simulations, which show a very reasonable behavior, have been simulated and presented. High performance could be achieved when calculating the backscatter data. The latter could be used in future work to predict or improve OTDR measurements.

Acknowledgments

The work described in this paper was supported by the grant Optika^2 from the Bayerisches Staatsministerium für Bildung und Kultus, Wissenschaft und Kunst. The authors gratefully acknowledge the support of the graduate study group Fiber Optic Transmission and Sensing (FiTS).

References

1. M. A. Losada, J. Mateo, J. J. Martínez, and A. López, "Si-pof frequency response obtained by solving the power flow equation," in *17th Intl. Conf. on Plastic Optical Fibres and Applications*, 2008.
2. D. Gloge, "Optical power flow in multimode fibers," *Bell System Technical Journal*, vol. 51, no. 8, pp. 1767–1783, 1972.
3. F. Breyer, "Multilevel transmission and equalization for polymer optical fiber systems," Ph.D. dissertation, Universität München, 2010.
4. B. Drljača, A. Djordjevich, and S. Savović, "Frequency response in step-index plastic optical fibers obtained by numerical solution of the time-dependent power flow equation," *Optics & Laser Technology*, vol. 44, no. 6, pp. 1808–1812, 2012.
5. D. Gloge, "Impulse response of clad optical multimode fibers," *Bell System Technical Journal*, vol. 52, no. 6, pp. 801–816, 1973.
6. A. W. Snyder and J. D. Love, *Optical waveguide theory.* Springer Science & Business Media, 2012.
7. O. Föllinger and M. Kluwe, *Laplace-und Fourier-Transformation.* Elitera Berlin, 1977.
8. O. Ziemann, J. Krauser, P. E. Zamzow, and W. Daum, *POF-Handbuch: optische Kurzstrecken-Übertragungssysteme.* Springer-Verlag, 2007.
9. C.-A. Bunge, R. Kruglov, and H. Poisel, "Rayleigh and Mie scattering in Polymer Optical Fibers," *Lightwave Technology, Journal of*, vol. 24, no. 8, pp. 3137–3146, 2006.
10. C. Emslie, "Polymer optical fibres," *Journal of materials science*, vol. 23, no. 7, pp. 2281–2293, 1988.
11. C. Gao and G. Farrell, "Modelling of rayleigh backscattering in plastic optical fiber," in *High Frequency Postgraduate Student Colloquium, 2003.* IEEE, 2003, pp. 14–17.
12. T. A. C. Flipsen, *Design, synthesis and properties of new materials based on densely crosslinked polymers for polymer optical fiber and amplifier applications.* Rijksuniversiteit

Groningen, 2000.
13. J. Brandrup, E. H. Immergut, E. A. Grulke, A. Abe, and D. R. Bloch, *Polymer handbook*. Wiley New York, 1999, vol. 89.
14. N. Tanio and Y. Koike, "Estimate of light scattering loss of amorphous polymer glass from its molecular structure," *Japanese journal of applied physics*, vol. 36, no. 2R, p. 743, 1997.
15. S. D. Personick, "Photon probe an optical-fiber time-domain reflectometer," *Bell System Technical Journal*, vol. 56, no. 3, pp. 355–366, 1977.
16. M. Mundus, J. Hohl-Ebinger, and W. Warta, "Estimation of angle-dependent mode coupling and attenuation in step-index plastic optical fibers from impulse responses," *Optics express*, vol. 21, no. 14, pp. 17 077–17 088, 2013.

III

Special fibre types

- tubular optical fibres
- active optical fibres

Modeling of Tubular Optical Fibers using the Example of an Optical-electrical Combination Conductor

B. Lustermann,[1,*] M. Viehmann,[1] E. Manske,[2]

[1] Nordhausen University of Applied Sciences, Weinberghof 4, 99734 Nordhausen, Germany.

[2] Ilmenau University of Technology, PO Box 100 565, 98684 Ilmenau, Germany.

*Corresponding author: lustermann@fh-nordhausen.de

The article deals with the modeling of tubular polymer optical fibers. These occur in the patented optical-electrical conductor combination system as light-conducting wrapping of an electrical cord and serve as a novel tool for the detection of arc faults. The wrapping represents a tubular step index waveguide consisting of 3 polymer layers: an inner and an outer cladding as well as a light-guiding core layer. To optimize the properties of the fiber, in particular the layer thicknesses, the influence of damping mechanisms caused by bending, core-cladding surface roughness and volume scattering losses are investigated and put into a model which is simple and flexible to parameterize. In doing so, the most difficult part turns out to be the modeling of surface-scattering at the core-cladding interface. In this paper the developed model is described and corresponding results are shown using the example of a silicone-FEP material combination.

1. Introduction

The rapid development of e-mobility leads to a series of new problems in the automotive industry. In particular, semi- and fully hybrid drive systems for motor- and railed vehicles require the use of high voltage onboard power supply. In such high voltage systems the risk of arc faults increases vastly. Serious damages to the entire vehicle and a high risk to people are associated with it. Therefore, an essential goal is the complete monitoring of the high voltage cables against arc faults. If it was possible to replace the normal insulation of an electrical cable by a rotationally symmetric optical waveguide, occurring arcs could be easily detected and so serious damage could be prevented.

The basic idea, patented by Viehmann [1], is to couple the light from the arc fault at the place of its origin into an optical waveguide to transmit this information to the optical detectors. In this way, after an arc fault has occurred, the circuit can be interrupted in the shortest possible time. The geometry of such a waveguide is necessarily tubular, that means the cross section is a circular ring with the electrical cable in the middle. At the same time the tubular optical waveguide forms the insulating layer of the electrical cable, thus giving the possibility to monitor the insulation with a pulsed or continuous light for possible mechanical damage or aging.

In order to transport the light of the arc fault along the cable, guiding by means of total reflection similar to conventional optical fibers is necessary. This requires coating of the light-guiding tube, the core, at both sides by a low reflective index cladding layer. The inner cladding layer separates the electrical conductor from the core and reduces the absorption of light by the electrical conductor, whereas the outer cladding layer delimits the core from the environment, see Fig. 1.

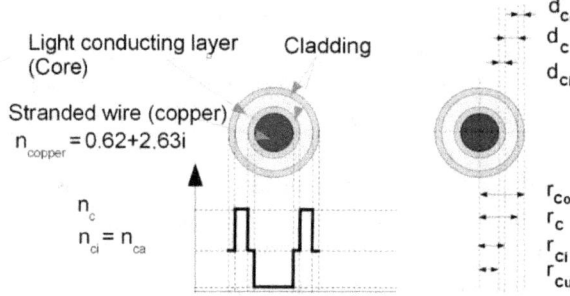

Fig. 1: Cross section of the optical-electrical conductor.

For the configuration of tubular light-guiding coating for electrical cables and for the design of the optical transmitter and receiver units, a number of new tube- and application specific questions arise. In particular, the influence of the optical material properties of the interface between core and cladding layers and the layer thicknesses have to be examined. Moreover, the use of such a cable in onboard power supplies imposes high mechanical and thermal requirements. In particular, the necessary flexibility makes the use of glass for this application impossible. Therefore, the focus is on the investigation of a polymer optical waveguide. This article describes the modeling method of such a tubular polymer waveguide as well as necessary roughness measurements of the core-cladding interface. Since the typical size of the waveguide is in the mm region, the modeling can be performed using ray-tracing. To describe the core-cladding interface roughness a scatter model, based on the Rayleigh-Rice Theory, is implemented and forms the focus of the modeling work.

2. Modeling of tubular optical waveguides

2.A. Materials

The results of material investigations of various polymers with respect to their suitability as optical fiber material for arc detection are summarized in the research report [2]. In particular, thermal requirements are the decisive criterion for the selection of cable material for use in a vehicle onboard power supply. Silicone elastomers as core material form the only group of materials which meets the required thermal limits while maintaining a high transparency. However, the special problem of the transparent variants of this material group lies in the production of the fibers. Since the starting materials are liquid, an extrusion is not possible. At present the production of such fibers takes place by homogeneous mixing, deaeration and vulcanization. Additionally, the centering of the copper wire turns out to be difficult. In our studies, Fluorpolymer (FEP) was used as cladding material. A similar combination of materials has been patented and studied by Zeidler for the production of conventional step-index optical fibers [3,4]. The results of his attenuation measurements are consistent with the simulation results using the fiber model which is described in the following.

2.B. Modeling of surface roughness

When simulating the scattering processes in a polymer optical fiber, it is important to put the focus on the modeling of the roughness of the core-cladding interface. Investigations of the influence of the interface roughness on the damping process have already been described in numerous publications [5,6]. Normally, after applying the cladding layer, a metrological determination of the interfacial structure with conventional surface roughness-measurement instruments is impossible. However, in the particular case of silicone-FEP-fibers, the low surface tension of both materials and the high value of ultimate elongation of silicone allow a non-destructive, subsequent removal of the cladding, so that the pure silicone core is available for metrological studies of roughness.

The scattering properties of a surface can be characterized by a probability distribution, indicating the distraction of impinging radiation from a certain direction. In literature the *Bidirectional Scatter Distribution Function* (BSDF) has been established [7]. The BSDF is defined as the ratio of the scattered intensity (I_{sc}) into a particular direction to the incident intensity I_i (θ and φ are polar and azimuthal angle, the index i stands for incoming and sc for scattered light):

$$BDSF(\theta_i, \varphi_i; \theta_{sc}, \varphi_{sc}) \, d\Omega_{sc} = \frac{dI_{sc}(\theta_i, \varphi_i; \theta_{sc}, \varphi_{sc})}{\cos\theta_{sc} I_i(\theta_i, \varphi_i)} \qquad (1)$$

In the limit of small roughness ($\ll \lambda$) the classical Rayleigh-Rice theory can be used. It provides a direct connection between the BSDF and the spectral density of the surface [8]:

$$BDSF(\theta_i, \varphi_i; \theta_{sc}, \varphi_{sc}) \, d\Omega_{sc} = \frac{4\pi^2 \Delta n^2}{\lambda^4} \cos\theta_i \cos\theta_{sc} Q(\theta_i, \theta_{sc}) S_{PSD}(f_x, f_y) \qquad (2)$$

Here, S_{PSD} is the 2D-power spectral density of the surface, Δn twice the refractive index of the core material. The factor Q depends on the reflection and transmission characteristics of the surface neglecting the roughness, whereas the spatial frequencies f_x and f_y depend on the incoming and scattered directions according to diffraction grating equations. In this way for each incoming ray the distribution of scattered light can be determined.

Metrologically, the surface is analyzed by straight scans. Therefore, the height profile is initially a set to 1D height profiles $z(y)$ from which the one-dimensional power spectral density (1D-PSD) can be calculated by use of Fourier transformation. Finally, in case of isotropic surfaces the two-dimensional (2D) PSD can be calculated from the 1D-PSD by means of an integral transform, see [7, Eq. (2-46)].

2.C. Experimental

The determination of the surface height profile can be carried out with different roughness measurement devices. Silicone polymer surfaces are very soft and therefore difficult to scan with tactile stylus instruments. These typically have a diamond probe with a 2 μm tip radius and a relatively large measuring force of more than 100 μN. Therefore, we used the laser scanning unit of the Nanopositioning & Nanomeasuring Machine NMM-1 of the Ilmenau University of Technology, see Fig. 2 (left), and the Laser Scanning Microscope Keyence VK-X100. The applied optical scanning found its metrological limit in the lateral direction by the minimum amount of available laser spot diameter of about 1 μm. In order to examine smaller surface structures, AFM (Atomic Force Microscopy) measurements with the NMM1 were also carried out in semi-contact mode (tip mode), see Fig. 2 (right).

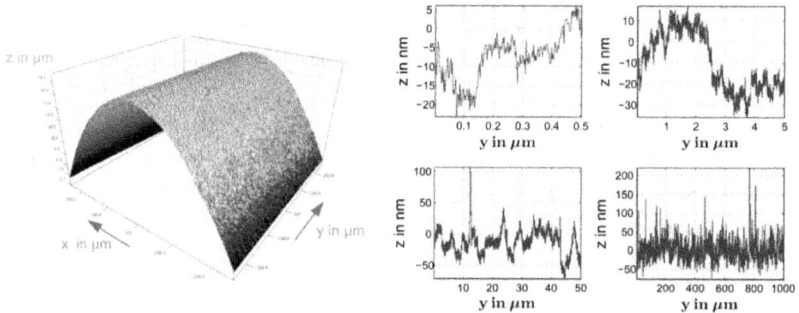

Fig. 2: NMM-1profile scans; left: 2D scan (500 μm × 500 μm) with laser focus sensor; right: 1D scan (1 mm) with AFM-sensor in semi-contact mode.

The sample step size was $\Delta y = 5\,\text{nm}$, the sample length 1 mm. One clearly sees that on small scales there is a very small roughness of a few nm, on larger scales of some 10 nm up to individual peaks of 100 nm...200 nm (probably due to impurities in the sample). The observed rms roughness at different measuring lengths confirms the assumption that the roughness parameters depend on the size of the analyzed surface segment and are smaller on smaller scales. Over the entire range an rms roughness of 23.8 nm was found.

Optical measurements performed with different profile measurement devices allow to obtain a lot of scan lines in a short time, which subsequently can be combined into a master PSD. This is done in Fig. 3, where the results of three different sets of measurements are shown together with the so-called ABC-Fit:

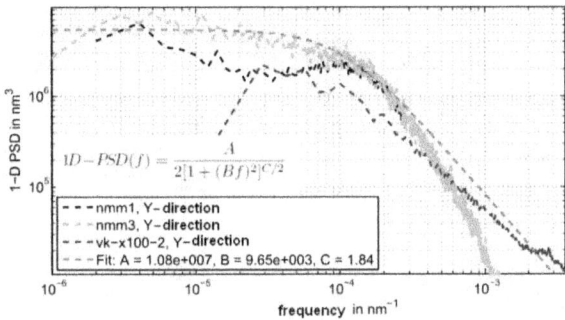

Fig. 3: Composite 1D -power spectral density (master PSD) measured along the fiber axes(Y-direction).

2.D. Simulation results

Using the modeling method described above, a scripted scatter model for the computation of the BSDF has been implemented in the ray-tracing software FRED [9] by means of which the optical behavior of arbitrarily composed waveguides can be examined, see Fig. 4 (left).

Fig. 4 (right) shows the simulation results exemplarily for a thickness of the outer cladding layer $d_{co} = 0.5$ mm, a radius of the electrical cable $r_{cu} = 0.5$ mm, a maximal angle of coupled light $\theta_{max} = 17.6°$, a refractive index of the core material $n_c = 1.41$ and of the cladding material $n_{ci} = n_{co} = 1.344$. The behavior of the curves leads to the conclusion that, under certain conditions, non-trivial optimal designs exist and can be found by the model. Their parameters are determined by the interaction between increasing scatter losses with decreasing core thickness on the one hand and increasing angle of acceptance if the hollowness of the waveguide increases on the other. In order to describe the hollowness of the fiber with a single parameter the tubularity $T = r_c^2/r_{ci}^2$ was established.

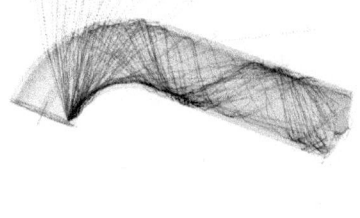 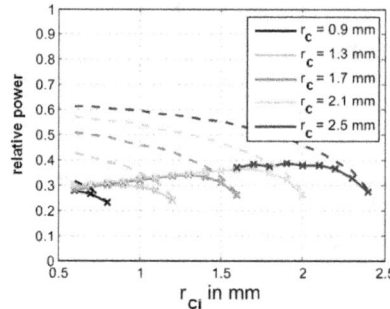

Fig. 4: Composed model of the optical-electrical combination conductor; right: simulation results of a bent optical-electrical combination conductor (2 m long, centered bending, bending radius $r_b = 16$ mm); dashed lines for smooth core-cladding-interface, continuous lines for an rms roughness $S_q = 30$ nm; r_c is the outer radius of the core layer, r_{ci} is the outer radius of the inner cladding layer.

3. Conclusion

A ray-tracing model of tubular optical waveguides including the effects of scattering on the core-cladding interface has been developed. With this flexible and easily configurable model it is possible to further optimize the design of the optical-electrical combination conductor. There is the possibility of an efficiency increase of up to 30% for straight fibers and of up to 10% for bent fibers for sufficiently smooth core-cladding interfaces if the respective layer design is optimal. With the model it is also possible to carry out bandwidth calculations and to optimize the arrangement of the transmitting and receiving units.

References

1. Viehmann, M.: Anordnung zur Überwachung elektrischer Einrichtungen auf das Entstehen von Störlichtbögen, Fachhochschule Nordhausen. 2005-04-28 (DE 103 42 370B3), patent.
2. Viehmann, M.; Kloß, C.: Grundlagen der Werkstofftechnik und der Fertigungstechnologie für einen optisch-elektrischen Kombinationsleiter, Forschungsprojekt DFG VI420/2-1 und 2-2, Juli 2009, research paper.
3. Zeidler, G.: Lichtleiter mit einem Kern aus Silikongummi und Verfahren zur Herstellung, 2002-04-18 (DE 101 54 945 A 1), patent.
4. Zeidler, G.: Elastomere Optische Fasern (EOF), in: ITG-FG Treffen Oldenburg (FTG15), March 25th/26th 2003.
5. Bierhoff, T.: Strahlenoptische Analyse der Wellenausbreitung und Modenkopplung in optisch hoch multi-modalen Wellenleitern, Universität Paderborn, dissertation, 2006.

6. Remillard, J. T.; Everson, M. P.; Weber, W. H.: Loss mechanisms in optical light pipes, in: Applied Optics 31 (1992), No. 34, pp. 7232-7241.
7. Stover, J.C.: Optical Scattering: measurement and analysis, 3rd ed. SPIE-Press, Bellingham, WA, USA, 2012.
8. Photon Engineering, LLC: Tutorial: Stray Light Short Course, September 2011, company brochure.
9. Photon Engineering, LLC: Optical Engineering Software FRED, www.photonengr.com, version: 2014, July 28th 2014.

Simulation of the Behavior of POFs Doped with Active Materials by Means of Ad-Hoc Finite-Difference Schemes

F. Jiménez,[1,*] J. Arrúe,[2] I. Ayesta,[1] M.A. Illarramendi,[3] J. Zubia[2]

[1] Dep. of Applied Mathematics, Faculty of Engineering of Bilbao, University of the Basque Country UPV/EHU; Alda. Urquijo s/n, E-48013 Bilbao, Spain.

[2] Dep. of Communications Engineering, Faculty of Engineering of Bilbao, University of the Basque Country UPV/EHU; Alda. Urquijo s/n, E-48013 Bilbao, Spain.

[3] Dep. of Applied Physics, Faculty of Engineering of Bilbao, University of the Basque Country UPV/EHU; Alda. Urquijo s/n, E-48013 Bilbao, Spain.

*Corresponding author: felipe.jimenez@ehu.es

In this chapter we share some of the expertise of the Applied Photonics Group of Bilbao in the numerical simulation of light propagation in POFs doped with fluorescent molecules. We describe the physical systems simulated, their modeling and governing equations, and some of the numerical techniques we use to solve them. Enough implementation details are provided for an interested researcher to be able to construct their own simulation software using similar concepts. During the course of the chapter we also try to give an idea of the kind of difficulties typically encountered in practice in the field of physical system simulation, and the kind of solutions that can be applied.

1. Introduction

Polymer optical fibers (POFs) present important advantages over glass fibers, even in the field of fiber lasers and amplifiers [1,2]. Apart from their well-known robustness and ease of handling, their lower manufacturing temperatures make it possible to use a wide range of materials as dopants, including organic dyes [3,4] and conjugated polymers [5,6]. This offers the possibility to obtain high-power broadband lasers in the visible using very short mirrorless fibers, thanks to the extraordinary large emission and absorption cross sections of such dyes in comparison to the conventional dopants of glass fibers like rare earths. Such POF lasers produce Amplified Spontaneous Emission (ASE) [7,8], which can be employed for the illumination of the inside of hollow objects, with the added advantage that low-coherence speckle-free light of high intensity can be generated inside very short lengths of POF if its properties are chosen adequately (length, dopant, concentration, pumping conditions, etc.).

Such light may be of interest, among others, in the field of Medicine or for interferometric gyroscopes [9]. POFs might also be employed for the generation of very intense light pulses in the infrared, such as those required in high-power all-fiber broadband super-continuum sources, which are even broader in spectrum than ASE sources, provided that they are doped adequately, e. g. with lanthanide complexes, because lanthanide ions do not undergo photo degradation, whereas very intense light pulses would rapidly degrade organic dyes.

In this chapter we will describe some of the techniques we use in the Applied Photonics Group of Bilbao for the numerical simulation of polymer optical fibers doped with active molecules. The chapter is organized as follows. Section 2 will give a brief description of the physical systems we simulate. Section 3 will describe their modeling and governing equations. Section 4 will give a general description of the numerical algorithms we have devised to solve them with different materials and excitation schemes. Some software architecture and implementation details are also provided. Finally, Section 5 will explain how we deal with some problems typically encountered in the numerical simulation of physical systems, like convergence and stability, memory management and computational cost.

2. Physical systems model

When a pump laser pulse of energy E_p is impinged onto a doped POF, either longitudinally or transversely, the fluorescent molecules of the dopant get excited. If the doped fiber is pumped with a high enough pump energy (above a threshold value) it can act as a mirrorless laser. Its emission (ASE) is much wider both spectrally and temporally than that of classical lasers. Alternatively, if a signal of energy E_s is added to the pump, the fiber may act as a tunable all-optical amplifier. Figure 1 illustrates these physical systems schematically.

The figure represents the simple case of a step-index (SI) fiber without cladding that is pumped longitudinally. In the absence of a signal, this system may act as a laser-like device. Otherwise a signal of wavelength λ_s, different and typically longer than the pump wavelength λ_p, may be introduced and optically amplified due to stimulated emissions inside the fiber. In the figure, z is the position along the fiber symmetry axis, with $z = 0$ on its left end, and n_1, n_2 are the refractive indices of the core and of the outer medium, respectively. Not only SI fibers, but also graded index (GI) fibers can be considered and used.

In Fig. 2 a pump laser of energy E_p emits light that impinges transversely onto the fiber in a perpendicular direction with respect to the fiber symmetry axis. This light illuminates a length z_e of fiber uniformly. A cylindrical lens can be employed to achieve this in an experimental set-up. In this system we set $z = 0$ on the left of the pumped volume, and we consider light propagation of the emission generated inside the fiber both in the $+z$ and $-z$ directions.

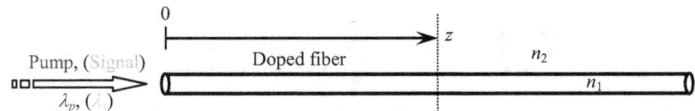

Fig. 1: Longitudinally-pumped doped fiber.

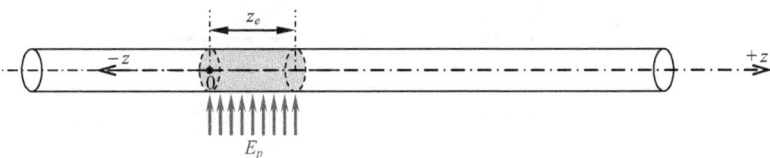

Fig. 2: Transversely-pumped doped fiber.

3. Modeling

In the case of dye-doped POFs, the fluorescent dopant molecules can be modeled as two-energy-state systems: the ground or unexcited electronic level S_0, and the excited electronic level S_1. Each level has many vibrational energy sublevels. We will call E_1 the lowest ground energy sublevel, and E_2 the lowest excited energy sublevel. These are represented in Fig. 3 as thicker horizontal lines.

From any energy sublevel, non-radiative vibrational relaxations, modeled as instantaneous, are assumed to occur, taking the dopant molecule to the corresponding lowest sublevel. Hence we assume that, at any given moment, every molecule is in either one of the two lowest energy sublevels, E_1 or E_2. We call $N_1(t,z)$ and $N_2(t,z)$ the respective concentrations in molecules per unit volume.

Molecule excitations take place from E_1 to any one of the sublevels of the excited state (depending on the energy of the excitation photon). The process is labeled as "Absorption" in the figure. Then the molecule will "instantaneously" relax to E_2. Emissions (both spontaneous and stimulated) take place from E_2 to any one of the sublevels of the ground state, again depending on the energy of the emitted photon. This is labeled as "Emission" in the figure. If the emission is stimulated, the emitted photon is of the same wavelength and direction as the photon that generated the emission. In contrast, if the emission is spontaneous, it can be assumed to be isotropic, and the wavelength of the emitted photon to be random with a probability density function proportional to the dopant's emission cross section as a function of wavelength λ.

Our model has three independent variables (time t, position z and wavelength λ) and tries to determine basically two unknown functions of them, namely the light power $P(t,z,\lambda)$ and

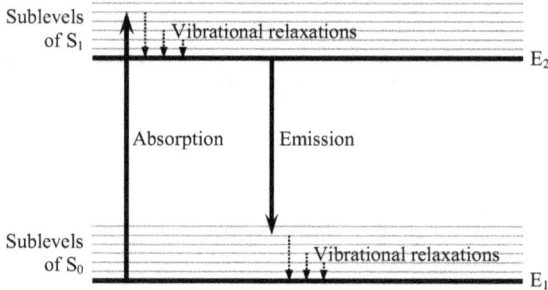

Fig. 3: Energy levels of the dopant molecules.

the concentration of excited dopant molecules $N_2(t,z)$. In some cases, like in side pumping, it may be desirable to also consider the light power propagating in the z direction $P(t,z,\lambda)$ in addition to P (which can be then called P^+). For the sake of simplicity we will give the explanations with only P, but the same ideas can be seamlessly applied to the propagation of P.

The rate equations are a set of partial differential equations that govern light propagation and excited dopant concentration in these systems, and are customarily written as follows:

$$\frac{\partial N_2}{\partial t} = \underbrace{-\frac{N_2}{\tau}}_{\text{spont. decay}} \underbrace{- \left(\frac{\sigma^e(\lambda)}{hc/\lambda A_{core}}\right) N_2 P \gamma}_{\text{stimulated decay}} + \underbrace{\left(\frac{\sigma^a(\lambda_p) N_1}{hc/\lambda A_{core}}\right) P_p \gamma}_{\text{excitation by the pump}} + \underbrace{\left(\frac{\sigma^a(\lambda)}{hc/\lambda A_{core}}\right) N_1 P \gamma}_{\text{excitation by } P \text{ (re-absorption)}} \quad (1)$$

$$\frac{\partial P_p}{\partial z} = \underbrace{-\sigma^a(\lambda_p) N_1 P_p \gamma}_{\text{material absorption of pump}} \underbrace{- \frac{1}{v_p}\frac{\partial P_p}{\partial t}}_{\text{propagation}} \quad (2)$$

$$\frac{\partial P}{\partial z} = \underbrace{\sigma^e(\lambda) N_2 P \gamma}_{\text{stimulated emission}} \underbrace{- \sigma^a(\lambda) N_1 P \gamma}_{\text{material absorption}} \underbrace{- \frac{1}{v}\frac{\partial P}{\partial t}}_{\text{propagation}} + \underbrace{\frac{N_2}{\tau}\left(h\frac{c}{\lambda}\right)\sigma_{sp}^e(\lambda)\beta A_{core}}_{\text{spontaneous emision}} \quad (3)$$

Equation (1) governs the rate of change of the concentration N_2 of excited molecules at each point z of the fiber and time t. In this equation, τ is the dopant molecule's mean lifetime; λ is the light wavelength; $\sigma^a(\lambda)$ and $\sigma^e(\lambda)$ are the absorption and emission cross sections at that wavelength, respectively; h is Planck's constant; c is the speed of light in vacuum; A_{core} is the fiber core's cross section area; λ_p is the wavelength of the pump; $N_1 = N - N_2$ (N being the total dopant concentration) is the concentration of non-excited molecules; and γ is an integral overlapping factor, equal to 1 in SI fibers and greater than 1 in GI ones, that takes into account that in GI fibers both the light power and the dopant concentrations are higher than average near the fiber symmetry axis, which increases the average rate of interactions.

Equation (2) rules the attenuated propagation of the longitudinally-pumped light power P_p, supposed monochromatic of wavelength λ_p — side pumping is modeled differently as will be explained in Sec. 4. In (2), v_p is the propagation speed of the pump light inside the fiber, and the rest of the symbols are as above. Finally, Eq. (3) governs the attenuated and amplified propagation of light of wavelength λ. In it, v is the propagation speed of that light inside the fiber; $\sigma_{sp}^e(\lambda)$ is the spontaneous emission cross section (normalized so that its integral over all is equal to 1), and β is the fraction of spontaneously emitted photons that lie in guided directions. Spontaneous emissions are isotropic, so in the case of SI fibers β can be simply calculated as a solid angle. In the case of GI fibers, an average value can be calculated.

Some of the descriptions above have been admittedly oversimplified, but the model and its simplifying assumptions can be considered as adequate inasmuch as the numerical simulations adequately reflect and predict the macroscopic behavior of the active POFs considered. It is also worth noting that the introduction of the spectral dependency of the absorption and emission cross sections, $\sigma^a(\lambda)$ and $\sigma^e(\lambda)$, is what allows a two-energy-state model to simulate spectral effects such as red and blue shifts, narrowing or widening of light spectra, etc.

For the rate equations to have a unique solution, initial and boundary conditions must be added to them. The initial conditions are typically $P(0, z, \lambda) \equiv 0$ and $N_2(0, z) \equiv 0$, i.e., initially the system has no light power propagating and no dopant molecules in the excited state.

The boundary conditions depend on whether the pumping is longitudinal or transverse. In the first case a monochromatic, Gaussian-in-time pulse is often employed, so

$$P(t, 0, \lambda_p) = P_{max} \cdot e^{\frac{-(t-t_0)^2}{2\sigma^2}}, \qquad (4)$$

where P_{max} is the maximum pump power, t_0 is the instant when it is reached, and σ is the Gaussian bell's temporal standard deviation or effective width. Another useful case, if the pump is not pulsed, is making $P(t, 0, \lambda_p) = P_p =$ const. This can sometimes be used to test the model by comparing its results with the exact solution in well-known stationary states.

The case of transverse pumping is treated differently and will be discussed in subsection 4.D below.

4. Numerical resolution

To solve the rate equations numerically we have devised ad-hoc algorithms based on finite differences with semi-discretization. By ad-hoc we mean that the concrete physical system we are modeling and the phenomena taking place in it are always borne in mind when coding the simulation software. The idea of semi-discretization is described in subsection 4.B below.

4.A. Basic scheme

The first thing we do is to discretize the three independent variables t, z, λ. For t, discrete instants $t_i = (i-1)\delta_t$ ($i = 1, 2, \ldots$) are considered. Similarly for z, discrete fiber points $z_j = (j-1)\delta_z$ ($j = 1, 2, \ldots, n$) are considered, with $z_n = L$ the fiber length. The

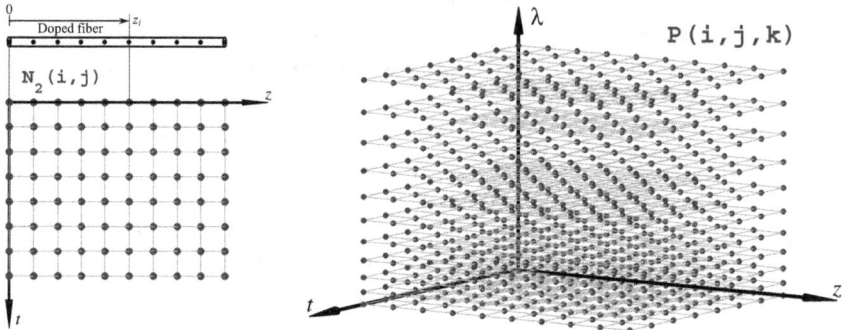

Fig. 4: Memory allocation for the discrete values of N_2 and P.

judicious choice of the values of δ_t and δ_z is important for the algorithm to converge without taking up too much computer memory and/or computational time. For the sake of simplicity we will proceed with the explanations as if both step sizes are constant, although this is not necessary.

The discretization of the wavelength λ is done a little bit differently. We consider a number of subintervals ("slots") of the whole range of wavelengths of interest. We call λ_k a wavelength representative of the k-th slot, like its midpoint. The width of these wavelength slots is typically not constant. It is often the case that the pump wavelength λ_p is the smallest one of all the wavelengths of interest (corresponding to the most energetic photons). In that case, the first wavelength subinterval is a very narrow one centered at λ_p, so $\lambda_1 = \lambda_p$. No photons of shorter wavelengths are considered to ever appear in the system, so the rest of the wavelength slots are located to the right of the first one. This method allows us to treat the propagation of the pump just like the propagation of light of any other wavelength, and Eq. (2) is treated as just a part of the discretized version of equation (3). In fact, emissions at λ_p do not need to be neglected as is customarily done – compare the terms in Eq. (2) and in (3)!

Once the discretization is ready, computer memory is allocated for the discrete values of N_2 and P. For $N_2(t,z)$, a two-dimensional matrix or array N2(i,j) is initialized. Similarly, for $P(t,z,\lambda)$, a three-dimensional matrix P(i,j,k) is used. This is illustrated in Fig. 4.

The value at any given array element N2(i,j) tries to approximate the excited molecule concentration at instant t_i and point z_j in the fiber. The value at any given element P(i,j,k) approximates the total light power propagating at that same instant and point with wavelengths contained in the k-th slot. Our program uses international-system units throughout, so the values of P(i,j,k) are in W.

Both matrices, N2 and P, are initialized as all-zeros. The initial condition $N_2(0,z) \equiv 0$ means that the first row of N2 (top row in Fig. 4, left) will always remain null. As pseudocode we can express this as N2(1,j)=0 for all j. The initial condition $P(0,z,) \equiv 0$ means that the first "hyper-row" of P will also remain null. In Fig. 4 (right) that corresponds to the vertical "plane" of dots at the back right of the figure. In pseudocode, P(1,j,k)=0 for all j and k.

Then the phenomena taking place in the system during each discrete time step δ_t are quantified and used to update N2 row by row and to update P "hyper-row by hyper-row". To code this it is useful to bear in mind the physical phenomenon accounted for by each term of the rate equations, as labeled in (1), (2), (3) above. This "ad-hoc" way of coding can sometimes take advantage of the knowledge we have about the physical system being simulated and it can show the way for improvements to the numerical algorithm.

The first term on the right-hand side of Eq. (1), $-N_2/\tau$, accounting for the decrease in N_2 produced by spontaneous decays, is calculated and multiplied by δ_t and hence converted into a discrete (finite) decrement. At the beginning of each time step the values of N_2 are found in a given row of the array N2. Hence the next row is copied from the previous one and then the calculated decrements due to spontaneous emissions are applied to it.

The second term, accounting for stimulated decays, is quantified in a similar way. $\sigma^e(\lambda)$ is computed for each wavelength slot as its integral value in the slot. The value is then multiplied by N_2 (taken from the corresponding position in the N2 array), by P (taken from the corresponding position in the P array), by (this integral value has been calculated and stored beforehand), and divided by h, c/k and A_{core}. Multiplying that by δ_t we obtain again a finite decrement of N_2 that is applied to the latest value at the corresponding new position of N2. Observe that each one of these decrements is usually small because it is the one produced by light power of wavelengths within just one wavelength slot; but after going through all the slots we will have a good approximation of the decreases in N_2, during δ_t, due to the decays stimulated by light of all the wavelengths considered (including the pump, which will often be the top contributor).

The other two terms in Eq. (1) are treated similarly, always updating the latest values of the elements in N2. As said above, the second-last term is actually treated as part of the last one in the general scheme, corresponding to the wavelength slot (typically the first one) where λ_p belongs.

The elements of P are updated in a similar way. For each wavelength slot, each term in Eq. (3) is approximated, multiplied by δ_z, and the finite increment or decrement so obtained applied to the latest appropriate value of P. Stimulated emissions (first term) contribute only to the value of P in the same wavelength slot being calculated, because the newly emitted photon is always of the same wavelength and direction as the incoming one. The next two terms are treated jointly as broadly described below in subsection 4.B. Finally, the last term in Eq. (3) is treated as follows. The overall rate of spontaneous emissions is calculated as N_2/τ just like we did to obtain the decrements of N_2 when discretizing the first term of Eq. (1)[4]. After multiplying by the constants shown in the term and by δ_z, a finite increment of P due to spontaneous emissions is obtained. And now the way to assign the generated light power

---

[4]Those values are actually stored and retrieved, rather than re-caclculated.

to the different wavelength slots is as follows: the spontaneous emission cross section $\sigma^e(\lambda)$ is normalized so its integral over all λ is equal to 1. The fraction of all spontaneously emitted photons with wavelengths within the k-th slot is then taken to be equal to the integral of the normalized emission cross section over that slot.

The description of the scheme we have provided has been necessarily oversimplified in some aspects, but a detailed description is out of the scope of this contribution. However, two more general remarks can still be made. Whenever centered finite differences can be used, the results tend to be more precise and stable than those of forward differences [11]. And it is sometimes also useful to interpolate from known values of N2 and P in the vicinity of the position and instant we are calculating.

4.B. Semi-discretization

In our experience, semi-discretization techniques can help with both the precision and the stability of the algorithm. In our case by semi-discretization we mean that a smooth, exact solution to some differential equation can be sometimes applied within a discrete time step δ_t. A clear example can be seen by considering the consecutive terms for material absorption and propagation in Eq. (2) or (3). If those were the only phenomena taking place (constantly attenuated propagation), the exact solution for P would be a well-known decreasing exponential of z. Hence, instead of approximating $\frac{\partial P}{\partial t}$ by means of finite differences (which is not very precise and even introduces elements of numerical instability[5]), the joint contribution of attenuation and propagation into P can be approximated by using an appropriate decreasing exponential. Of course this must be done carefully, and more sophisticated versions of this same idea can also be implemented by calculating the exact solution to some other reasonable differential equations that take into account more than these two terms.

4.C. Different dopants

The software architecture is designed so that the same simulation kernel can handle different dopants by simply invoking different functions that calculate their absorption and emission cross sections at any wavelength λ. The data to construct these auxiliary functions are sometimes taken from the literature, either numerically or carefully scanned, or sometimes also measured by us in our laboratory. Then the function interpolates those data to obtain the value at any wavelength, which is necessary in order to carry out further manipulations like numerical integrations. For instance, for the cases of the organic dyes rhodamine B and PFO, the absorption and emission cross sections have the aspect shown in Fig. 5, generated with our auxiliary functions for [4] and reproduced here.

The two names (identifiers or "handles") of the functions that calculate these two cross sections at any λ are passed to the main simulation program as just another two input arguments.

---

[5]A constant decrease at the initial rate can easily produce negative values of P or of N_2 if δ_t is too large, while a decreasing exponential will never do that.

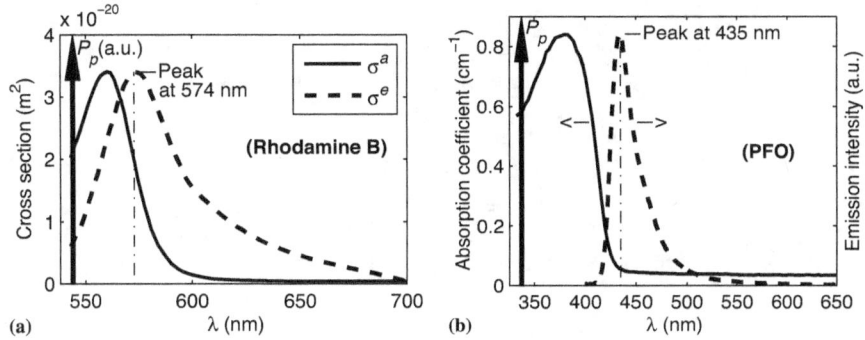

Fig. 5: Absorption and emission cross sections of Rhodamine B (a) and PFO (b).

4.D. Different pumping set-ups and boundary conditions

Other arguments are also passed to inform the main program about the pumping conditions (whether it is longitudinal or lateral, and with what parameters). Since many of the numerical steps of the simulation are common for different dopants and pumping schemes, it makes sense that one only program can handle all these cases.

Longitudinal pumping translates very easily into neat boundary conditions of direct application in the numerical scheme described above. For a Gaussian-in-time pulsed pump, Eq. (4) can be just evaluated at the instants $t = t_i$ of the temporal discretization, and the values obtained are then placed into P(i,1,1) (the second index being 1 to correspond to $z = 0$, and the third index being also 1 if, as is often the case, the pump is monochromatic and the first wavelength slot is the one including λ_p). For a continuous pump, P(i,1,1) is filled with the constant P_p.

On the other hand, side or transversal pumping does not translate into such a neat set of boundary conditions, because transversely-impinged pump light is perpendicular to the fiber symmetry axis and therefore never lies in guided directions. Figure 6 is a cross-sectional representation of several paths that pump rays might follow. It is only schematic: the trajectories of rays in a GI fiber are depicted as straight lines, the figure does not show scattering or possibly appearing skewed rays, etc. However, the figure should be enough to show that an average length $D \approx 2\rho$ of interaction between pump rays and dopant may be used as a first approximation, although a more realistic model might be achieved by using the ray-tracing method [10]. The absorbed pump power can then be approximated as $P_{abs}(t) = P_p[1 - \exp(\sigma^a(\lambda_p)N_1(t)D)]$ for $0 \leq z \leq z_e$. In other words, for every time step δ_t, an average attenuation coefficient α', which is proportional to $N_1 = N - N_2$ (taken from matrix N2 at each point and time), is applied over the average interaction length D to estimate the number of dopant molecules excited by the pump. Dividing this by the active volume, we

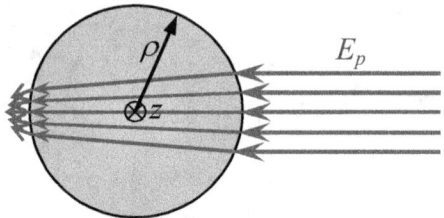

Fig. 6: Schematic representation of some of the ray paths in transverse pumping.

obtain the increments to be applied directly to the relevant positions of the array N2.

In our model, the only effect of side illumination is the excitation of dopant molecules inside the active volume (in green in Figs. 2 and 6). The spontaneous or stimulated emissions of those molecules will be the only contributors to $P(t, z, \lambda)$, and these processes are quantified by the common part of the program that works all the same whether the pump is longitudinal or transverse. Light scattering in guided directions, which are almost perpendicular to the incident rays, can be added to the model to simulate side illumination a bit more precisely, but we have not done this yet.

5. Stability, memory use, computational cost

Numerical instability is a phenomenon encountered in practice when running a wide variety of numerical algorithms. When a method is unstable, rounding and truncation errors propagate through the algorithm in an amplified or exponential manner, and the numerical results do not converge to the exact solution of the problem. In fact, in many cases, what the researcher actually gets is a set of totally useless numbers differing from the exact solution by many orders of magnitude. Even very simple finite-difference schemes are not immune to numerical instability [11].

The stability of a method typically depends not only on how well the method has been designed, but also on the parameters chosen to run it. Given its real-life pervasiveness, instability has been studied theoretically with attention. Still, without hands-on experience, it is often very difficult to know if an even moderately complex numerical scheme will become unstable or not when run with a given set of parameters. In the case of finite-difference schemes, the choice of the discretization step sizes is often critical. The shorter those steps are (while also sometimes respecting certain relationships between them), the better for stability and also for precision, if stability is achieved.

However, choosing very small step sizes greatly increases computational cost and computer memory usage, specially when the number of dimensions (variables) to be discretized is more than one or two. For example, if we chose a temporal step size δ_t small enough that 10^3 time steps have to be computed, a spatial step size δ_z small enough to consider 10^3 discrete points z_j in the fiber, and a wavelength discretization with 10^3 subintervals in order to obtain very smooth spectra, the number of array elements P(i,j,k) to be calculated would be $(10^3)^3 = 10^9$, rendering the method unacceptably slow even in a fast computer. Also bear in mind that just a single figure for a research paper often needs dozens of simulations to obtain the points of each curve in the figure, so one may not afford to wait for even one hour for a simulation to finish. Long runtimes are only partially solved by launching nightly batch jobs, because the process must often be very interactive.

Memory requirements could also become an issue. In the example above, if one tries to initialize P = zeros(1000,1000,1000), one will often get an "Out of memory" or similar error. This can happen even if the computer has a lot of available memory from the hard disk to swap because of hard-coded limits imposed by the numerical software used. However, this problem is relatively easy to palliate. For example, if one really needed such a fine discretization for stability reasons, one could advance in those tiny time steps δ_t, but store the results obtained only in one out of ten of them. That would turn the initialization command above into P = zeros(100,1000,1000), which may be enough to make a difference with regard to memory allocation. And if this is not enough, the same could be done with the other two independent variables.

However, none of this helps with the computational cost, because, even if one does not store all of them, one would still need to calculate more than 10^9 values. When stability requirements impose step sizes that are too small to finish in reasonable computing times, three main things can be done:

- Improve the numerical algorithms in terms of stability (because it imposes hard limits on the step sizes that can be used) and, less importantly, in terms of precision. Some techniques in that sense have already been mentioned, like trying to use centered finite differences or semidiscretization techniques when possible. As far as precision is concerned, Richardson's extrapolation [11] is another fairly general-purpose idea that is easy to implement.

- Change the hardware. Buy or rent a supercomputer. Use more than one computer simultaneously (if the algorithm can be parallelized, which is often not the case).

- Change the software. Specifically, change the programming language.

Most of our simulation work has traditionally been done with Matlab®(for interactivity) and C (for speed). Interpreted dynamic languages, like Matlab or Python, are great for user interactivity, fast graphical representations and prototyping; but for the innermost loops where the code needs to run fast, their speed is often just not enough. On the other hand, compiled static languages like C or Fortran are basically as fast as it gets (save writing ad-hoc machine code, which is not an option for us), but the interactivity is lost. This is sometimes called the Ousterhout dichotomy [12]. What many researchers do in practice is develop their ideas in an interactive scripting language and then port the parts that must be fast to a statically compiled one. But this solution is not optimal—one would prefer to use the same language for everything.

Some modern languages are trying to overcome this dichotomy, and Julia [13], whose development started in 2009 at MIT, looks promising. Julia is a general-purpose language specialized in technical computing. Its performance is very high, with speeds comparable to those of C or Fortran (sometimes even faster). The language itself is a very high-level and expressive one, with a syntax quite similar to Matlab's for instance (although the internal language design is necessarily different in order to be performant). It is possible to write very readable, compact and elegant code in Julia. The ability to provide both interactivity and performance lies in some of its basic design choices, including the ones that allow it to be efficiently compiled just-in-time. Last but not least, Julia is free (under a MIT license). For these reasons, we are presently porting much of the code described above from Matlab to Julia.

6. Conclusions

In this chapter we have tried to share some of the experience of the Applied Photonics Group of Bilbao in the numerical simulation of light propagation in POFs doped with fluorescent materials. After introducing the physical systems that we simulate, we have described the type of numerical scheme we use to solve their governing equations. Even if a detailed explanation of every line of code is out of the scope of this text, we hope to have provided enough architecture and implementation details for an interested researcher to be able to write their own simulation software along the same ideas.

Writing the numerical software "ad-hoc", i. e. with the specific physical systems and processes being simulated always in mind, and not just for an abstract set of differential equations detached from the concrete systems they govern, is, in our opinion, useful in order to find solutions to the practical problems typically encountered in the field of physical system simulation, like numerical instability, convergence, or computational cost.

Acknowledgements

This work has been funded in part by the Fondo Europeo de Desarrollo Regional (FEDER); by the Ministerio de Economía y Competitividad of Spain under project TEC2012-37983-C03-01; by the Gobierno Vasco / Eusko Jaurlaritza under projects IT664-13 and ETORTEK14/13; and by the University of the Basque Country UPV/EHU under programs UFI11/16 and US13/09.

References

1. J. Arrue, F. Jiménez, M.A. Illarramendi, J. Zubia, I. Ayesta, I. Bikandi and A. Berganza, "Computational analysis of the power spectral shifts and widths along dye-doped polymer optical fibers," IEEE Photon. J., vol. 2, no. 3, pp. 521-531, Jun. 2010.

2. I. Ayesta, J. Arrue, F. Jiménez, M. Illarramendi and J. Zubia, "Computational analysis of the amplification features of active plastic optical fibers," Phys. Status Solidi (a), vol. 208, no. 8, pp. 1845-1848, 2011.

3. I. Ayesta, J. Arrue, F. Jiménez, M.A. Illarramendi and J. Zubia, "Analysis of the emission features in graded-index polymer optical fiber amplifiers," J. Lightw. Technol., vol. 29, no. 17, pp. 2629-2635, Sep. 2011.

4. J. Arrue, F. Jiménez, I. Ayesta, M.A. Illarramendi and J. Zubia, "Polymer-optical-fiber lasers and amplifiers doped with organic dyes," Polymers, vol. 3, no. 3, pp. 1162-1180, 2011.

5. I. Ayesta, M.A. Illarramendi, J. Arrue, F. Jiménez, J. Zubia, I. Bikandi, J.M. Ugartemendia and J.R. Sarasua, "Luminescence study of polymer optical fibers doped with conjugated polymers," J. Lightw. Technol., vol. 30, no. 21, pp. 3367-3375, Nov. 2012.

6. I. Bikandi, M.A. Illarramendi, J. Zubia, J. Arrue, and F. Jiménez, "Side-illumination fluorescence critical angle: Theory and application to F8BT-doped polymer optical fibers," Opt. Exp., vol. 20, no. 4, pp. 4630-4644, Feb. 2012.

7. M.A. Illarramendi, J. Arrue, I. Ayesta, F. Jiménez, J. Zubia, I. Bikandi, A. Tagaya and Y. Koike, "Amplified spontaneous emission in graded-index polymer optical fibers: theory and experiment," Opt. Exp., vol. 21, no. 20, pp. 24254-24266, Oct. 2013.

8. J. Arrue, M.A. Illarramendi, I. Ayesta, F. Jiménez, J. Zubia, A. Tagaya and Y. Koike, "Laser-like performance of side-pumped dye-doped polymer optical fibers," IEEE Photon. J., vol. 7, no. 2, pp. 1-11, Apr. 2015.

9. J. Arrue, I. Parola, I. Ayesta, F. Jiménez, M.A. Illarramendi, J. Zubia, A. Tagaya and Y. Koike, "Theoretical and Experimental Analysis of Polymer Optical Fibers Working as Broadband Lasers," in Proceedings of the 24th International Conference on Plastic Optical Fibers (POF 2015) in Nürnberg, Germany, 2015.

10. F. Jiménez, J. Arrue, G. Aldabaldetreku and J. Zubia, "Numerical simulation of light propagation in plastic optical fibres of arbitrary 3D geometry," WSEAS Trans. Math, vol. 3, pp. 824-829, 2004.

11. R.L. Burden, J.D. Faires, Numerical Analysis, Sixth Edition, Brooks/Cole Publishing Company, 1997.

12. Technicalities: interactive scientific computing..., `http://graydon2.dreamwidth.org/3186.html`.

13. The Julia programming language home page, `http://julialang.org/`.

IV

Sensor applications

- gamma-radiation induced effects in perfluorinated POFs
- POF sensors in energy, oil and biotechnology areas
- optical power monitor with SI-POFs

Gamma Radiation Induced Effects in Perfluorinated Polymer Optical Fibers for Sensing Applications

P. Stajanca,[1,*] D. Sporea,[2] L. Mihai,[2] D. Negut,[3] M. Schukar,[1] K. Krebber[1]

[1] *BAM Federal Institute for Materials Research and Testing, Unter den Eichen 87, 12205 Berlin, Germany.*

[2] *National Institute for Laser Plasma and Radiation Physics, Atomistilor St. 409, RO-077125 Magurele, Romania.*

[3] *"Horia Hulubei" National Institute of Physics and Nuclear Engineering, Reactorului St. 30, Magurele, Romania.*

Corresponding author: pavol.stajanca@bam.de

Radiation induced attenuation in perfluorinated polymer optical fiber was investigated as a potential mean for online, in situ gamma radiation monitoring. Spectral dependence of radiation induced attenuation was measured in the visible and near infrared range. Attenuation spectrum exhibits sharp increase both towards the ultraviolet and infrared side of the spectrum with 3 distinct peaks around 1400 nm. The fiber sensitivity to gamma radiation was further evaluated at two commonly used IR wavelengths (1310 nm and 1550 nm) by pre- and post-irradiation optical time-domain reflectometry measurements. The examined fiber is sensitive up to 100 kGy of total absorbed dose with only minor saturation towards the large dose values. The influence of gamma radiation on the fiber mechanical properties was studied as well. Up to the maximal dose of 100 kGy, no significant deterioration of fiber strength and elasticity was observed.

1. Introduction

In the last decades, ionizing radiation is finding new and new application in various fields including sterilization, material processing or medicine. Irradiation process needs to be controlled and monitored, and therefore also demand for new, improved dosimeters is rising. Among the broad variety of available radiation dosimeters, optical fiber-based sensors offer a unique possibility of remote and real time monitoring. In addition, small dimensions, high flexibility and electromagnetic immunity allow for monitoring in confined or harsh environments [1]. A number of different optical fiber dosimetry techniques have been developed relying on various sensing mechanisms [2,3]. One of the most straightforward techniques is based on a simple fiber attenuation monitoring. Ionizing radiation degrades the glass material of the optical fiber which manifests itself as an increase of the fiber losses. The degree of radiation induced attenuation (RIA) can be correlated with the absorbed dose [4,5]. Induced attenuation is wavelength dependent and determined by the glass composition [2]. Radiation sensitivity of pure silica glass is relatively low and different dopants must be used in order to achieve required sensitivity [6,7]. RIA in glass optical fibers has been investigated extensively with earliest deployments of RIA-based optical fiber dosimeters (OFD) dating back to 1970's [8].

In the last years, significant progress in the development of polymer optical fibers (POFs) has been achieved, fueled mostly by the growth of telecommunication industry [9]. For applications that do not require long transmission range, POFs may represent more suitable and cheaper alternative. Large diameter multi-mode (MM) POFs are very robust but still flexible, offer more economical means of connectorization and can therefore yield low-cost user-friendly systems. In addition, polymer fibers don't break in the brittle manner, have better biocompatibility and are generally more acceptable for medical applications. With regard to radiation dosimetry, optical fibers based on polymer material may offer higher inherent radiation sensitivity without need for additional doping as in case of their silica counterparts [3]. Recently, gamma radiation OFD using commercial MM polymethyl methacrylate (PMMA) fiber was demonstrated [10–12]. The fiber exhibited good sensitivity as high as $0.6\,\text{dBm}^{-1}/\text{kGy}$ and wavelength dependent sensing range of 0.03-45 kGy, which is broader than for any present commercially available sensor [12]. However, operation of PMMA-based OFD is limited to short fiber lengths and visible spectral region due to the high inherent material optical absorption of polymer materials. Compared to POFs based on common optical polymers such as PMMA or polycarbonate, perfluorinated polymer optical fibers (pPOFs) can offer relatively low-loss transmission in much broader spectral region covering visible (VIS) and near infrared (NIR) part of the spectrum [13]. Operation in low-loss NIR region, where more developed optoelectronic components are available, may contribute to improvement of the sensor performance. In addition, possibility of long transmission ranges allows also for distributed radiation sensing that might be interesting for certain application, e. g. monitoring of accelerator beam-lines [14]. In this work, we present preliminary study into the radiation induced effects in perfluorinated polymer optical fibers. In order to assess the feasibility of pPOF-based optical fiber dosimeter, we investigate the gamma radiation induced attenuation in a commercial MM pPOF with focus on radiation sensitivity in the NIR spectral region.

2. Fiber irradiation and characterization

All the presented experiments were performed on a commercial MM perfluorinated fiber GigaPOF-50SR form Chromis Fiberoptics. Fiber has an overall diameter of 490 μm and 50 μm graded-index core. For optical time-domain reflectometry (OTDR) measurements, five 20 m long samples were cut from the fiber. Both ends of the samples were polished and termi-

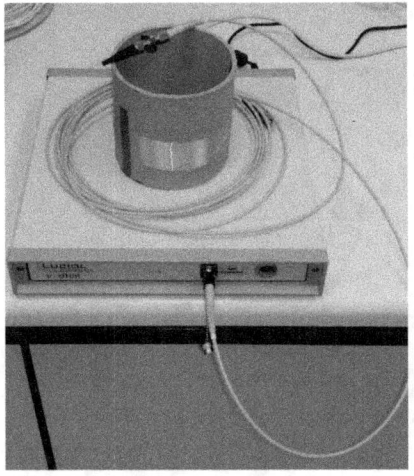

Fig. 1: Sample of investigated fiber spooled around the cardboard cylinder for OTDR measurements (left). In-house built tensile testing machine used for mechanical testing of investigated fiber (right).

nated with FC/APC connectors. In order to fit samples into round irradiation chamber, each sample was spooled around a separate cardboard cylinder with diameter of 10.5 cm (Fig. 1 (left)). Another set of 5 shorter, freely spooled samples without connectors were prepared to be irradiated for mechanical testing and spectral attenuation measurements.

The samples were irradiated with Co^{60} source at 5.7 kGy/h dose rate. Each samples was irradiated to different level of total dose. Dose levels of 1, 5, 20, 50 and 100 kGy were used. Fiber OTDR traces were measured from both sample ends before and after irradiation utilizing ν-OTDR from Luciol Instruments (Fig. 1 (left)). This OTDR unit allows for measurement at two different wavelengths (1310 nm and 1550 nm) in the NIR region. Spectral character of attenuation of virgin and irradiated fiber was measured by cutback method utilizing Yokogawa AQ4305 white light source and Advantes Q8347 optical spectrum analyzer.

In order to assess influence of gamma irradiation on the fiber mechanical properties, fiber strain test was performed on original non-irradiated, as well as irradiated samples. Strain testing was performed on an in-house built tensile testing machine (Fig. 1 (right)) The machine is composed of one static fiber holder with a load cell for the force measurement and one stepper-motor-driven traveling crosshead with linear displacement sensor for fiber straining. Samples with an approximate gauge length of 100 mm were strained at $66 \pm 1\%$/min. Sample initial diameter was measured by micrometer screw gauge at 3 different points along the strained length of the fiber. All presented results were averaged from minimum of 4 tested samples.

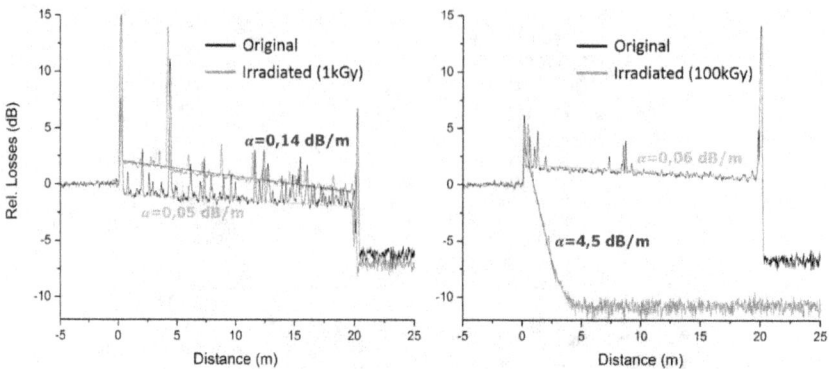

Fig. 2: Pre- and post- irradiation OTDR traces measured at 1310 nm for the pPOF sample irradiated to the total dose level of 1 kGy (left) and 100 kGy (right). Estimated attenuation coefficients are indicated.

3. Results

The investigated pPOF has one of the attenuation minima at 1310 nm. The fiber attenuation should be lower than 60 dB/km at this wavelength, as specified by manufacturer [15]. Pre- and post-irradiation OTDR traces of 2 fiber samples (irradiated to 1 kGy and 100 kGy) measured at this wavelength are depicted in Fig. 2. The backscatter signal has rather structured character with a number of reflection features, which is typical for this type of pPOF [16]. Nevertheless, an approximate attenuation coefficient α' is determined from a general slope of the fiber backscatter trace (Fig. 2). Determination of the fiber attenuation from OTDR trace for such short fiber samples has a limited accuracy. In addition, for samples subjected to higher total doses, radiation induced losses (RIA) are so high that signal reaches OTDR's detection limit even before the end of the fiber. This shortens the distance from which the attenuation is evaluated, which further increases uncertainty. Nevertheless, achieved accuracy is sufficient to qualitatively demonstrate the effects of gamma radiation on fiber transmission and preliminary assess their magnitude. For non-irradiated samples, attenuation values at the level of 0.05-0.06 dB/m were estimated, which is in agreement with manufacturer specifications. A subtle but measurable increase of fiber attenuation at this wavelength is already noticeable after irradiation to 1 kGy, which was the smallest total dose utilized in this study. Along with attenuation increase to 0.14 ± 0.01 dB/m, irradiation also appears to increase the fiber backscatter level. However, the increased backscatter level might be as well associated with the change of light launch conditions into the fiber and needs to be confirmed by further on-line measurements, where no change in the fiber inter-connections can be guaranteed. The attenuation gradually increases with rising dose level up to 4.5 ± 0.2 dB/m measured for the sample irradiated to the maximal applied dose of 100 kGy.

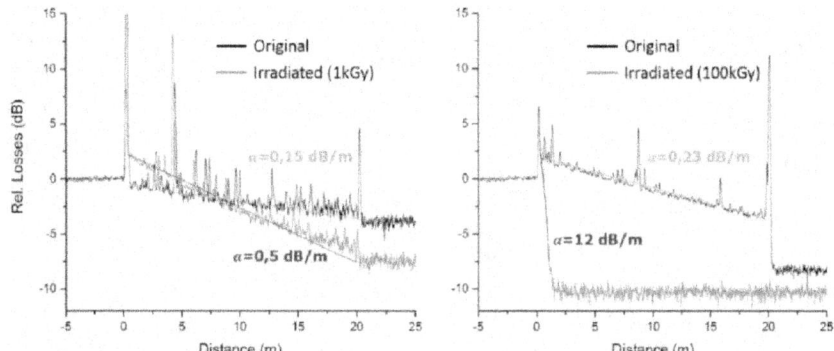

Fig. 3: Pre- and post- irradiation OTDR traces measured at 1550 nm for the pPOF sample irradiated to the total dose level of 1 kGy(left) and 100 kGy (right). Estimated attenuation coefficients are indicated.

Analogical behavior is observed also for the backscatter traces measured at 1550 nm (Fig. 3). According to our OTDR measurements, intrinsic fiber attenuation is more than 3 times higher at this wavelength, which would limit the utilizable fiber length. On the other hand, the radiation sensitivity is also higher at this wavelength with post-irradiation attenuation increasing from 0.5 ± 0.05 dBm measured after 1 kGy irradiation to 12 ± 1 dB/m recorded after 100 kGy irradiation.

The OTDR traces were measured from both fiber ends for every sample. The estimated attenuation coefficient values were then averaged from the bi-directional measurements. The RIA value for each sample was determined as the difference between post- and pre-irradiation attenuation estimated from the OTDR measurements. The RIA dependence on the total absorbed dose for both wavelengths is depicted in Fig. 4 (left).

The tested fiber is sensitive up to maximal applied dose of 100 kGy with gradual saturation towards the large dose values. Assuming the linear response for the small dose levels up to 5 kGy, the radiation sensitivity of 0.096 ± 0.006 dBm^{-1}/kGy at 1310 nm and 0.25 ± 0.05 dBm^{-1}/kGy at 1550 nm was estimated by linear regression. Even though this is several times lower that certain special highly sensitive glass fibers exhibit in this spectral range [7,17], it is much higher than standard Ge-doped silica fibers offer [7] and comparable to PMMA POF sensitivity in the VIS range [12]. Generally, the sensitivity of investigated pPOF at 1550 nm is almost 3 times higher than at 1310 nm. This is in agreement with the spectral attenuation measurement performed on the fiber irradiated to 5 kGy (Fig. 4 (right)). Comparing the spectral attenuation of non-irradiated and irradiated fiber, one can see that RIA has rather complex spectral shape. Similar as in case of glass optical fibers, radiation sensitivity rises dramatically towards the UV part of the spectrum. Presented results indicate that the radiation sensitivity

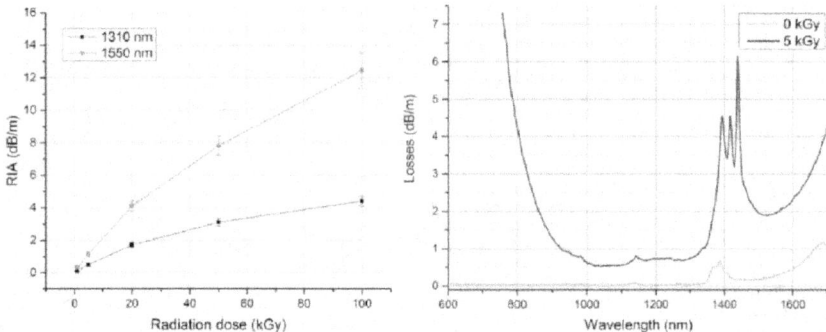

Fig. 4: Radiation induced attenuation (RIA) of investigated perfluorinated optical fiber as a function of total absorbed dose measured at two commonly used NIR wavelengths (left). Spectral attenuation of original non-irradiated fiber and sample irradiated to 5 kGy of total dose (right).

of pPOFs in the visible range can be much higher than for PMMA-based fibers [12]. Besides rapid increase towards UV, RIA increases significantly also on the IR side with three distinct peaks around 1400 nm. As demonstrated before, this allows for efficient radiation sensing in the NIR region, where fiber-optoelectronic components are generally more developed. The RIA spectrum has a minimum in the 1000-1100 nm region. The increase of attenuation on the blue side of the spectra in glass optical fibers is typically attributed to creation of electron-hole defects in the glass structure. These defects then act as electron traps forming the color centers. Analogically, increase of UV/VIS absorption in case of POFs can be explained by generation of free radicals by gamma radiation-induced scission of polymer chains. Alternatively, the increase of conjugation of polymer molecular structure can be responsible for observed behavior [18]. Detected IR absorption features are typical for higher overtones of C-H vibration modes [19]. This would indicate scission of C-F bonds in the perfluorinated molecule and creation of new C-H bonds instead. Another effect that may contribute to higher attenuation is increase of scattering as indicated by presented OTDR traces. However, further investigation is required to fully understand the mechanism of RIA increase in studied perfluorinated polymer optical fiber.

Besides optical properties, polymer structural changes caused by gamma radiation may also lead to significant change of fiber mechanical properties. For the fiber sensor to remain practical, fiber needs to retain its flexibility after irradiation. Therefore, we performed the fiber strain test for original non-irradiated and irradiated fiber samples. The effects of gamma radiation on the fiber mechanical performance are rather minor and no significant deterioration was observed even after maximal irradiation to 100 kGy (Fig. 5). The investigated fiber is very ductile, with the fiber ultimate strain reaching values around 90% before irradiation.

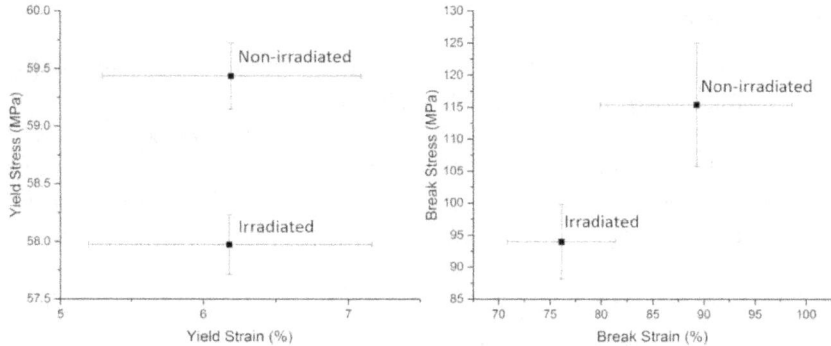

Fig. 5: Main mechanical parameters of original non-irradiated fiber and fiber irradiated to maximal utilized total dose of 100 kGy.

The mild decrease of fiber strength is apparent from lower yield and ultimate stress values. Also fiber ductility decreases with fiber ultimate strain dropping by roughly 15% after maximal irradiation. However, recorded changes are rather small and do not represent serious limitation for fiber mechanical performance.

4. Conclusion

Gamma radiation sensitivity of commercial multi-mode perfluorinated polymer optical fiber was investigated for the first time. Fiber sensitivity up to 100 kGy of total absorbed dose was demonstrated by OTDR measurements at two commonly used wavelengths in the near infrared region. Due to lower inherent losses, operation at 1310 nm allows for longer effective fiber lengths, which might be of interest for radiation distributed sensing. On the other hand, operation at 1550 nm offers almost 3 times higher sensitivity at the level of 0.25 dBm^{-1}/m, which is comparable with the sensitivity of PMMA fibers in the visible range. Spectral dependence of radiation induced attenuation was measured as well. Spectral shape of RIA exhibits rapid increase towards both UV and IR side of the spectrum with minimum between 1000-1100 nm and three distinct peaks around 1400 nm. The results obtained for VIS region indicate that pPOF may yield dosimeters with much higher sensitivity than PMMA-based fibers operated in this spectral range. On the other hand, operation in IR region can be more compelling due to availability of more advance fiber and optoelectronic components. Possible origins and mechanism of optical losses increase were discussed. However, more detailed study into this topic is required. Mechanical testing performed on irradiated fibers didn't reveal any significant deterioration of the fiber mechanical performance even after irradiation to 100 kGy which was the maximal total dose value utilized in this study.

Acknowledgements

The research leading to these results has received funding from the People Programme (Marie Curie Actions) of the European Union's Seventh Framework Programme FP7/2007 2013/ under REA grant agreement no 608382. The Romanian authors acknowledge the supported of the Romanian Executive Agency for Higher Education, Research, Development and Innovation Funding (UEFISCDI), under Grant 8/2012, project "Sensor Systems for Secure Operation of Critical Installations".

References

1. T. Shikama, K. Toh, S. Nagata and B. Tsuchiya, "Optical dosimetry for ionizing radiation fields by infrared radioluminescence," Measurement Science and Technology, vol. 17, no. 5, pp. 1103-6, 2006.
2. A. L. Huston, B. L. Justus, P. L. Falkstein, R. W. Miller, H. Ning and R. Altemus, "Remote optical fiber dosimetry," Nuclear Instruments and Methods in Physics B, vol. 184, pp. 55-67, 2001.
3. S. O'Keeffe, C. Fitzpatrick, E. Lewis and A. I. Al-Shamma'a, "A review of optical fibre radiation dosimeters," Sensor Review, vol. 28, no. 2, pp. 136-142, 2008.
4. K. Awaza, H. Kawazoe and M. Yamane, "Simultaneous generation of optical absorption bands at 5.14 and 0.452 eV in 9 $SiO_2:GeO_2$ glasses heated under an H_2 atmosphere," Journal of Applied Physics, vol. 68, no. 6, pp. 2713-18, 1990.
5. D. L. Griscom, "Optical Properties and Structure of Defects in Silica Glass," Journal of the Ceramic Society Japan, vol. 99, no. 10, pp. 923-42, 1991.
6. D. L. Griscom and E. J. Friebele, "Effects of ionizing radiation on amorphous insulators," Radiation Effects, vol. 65, pp. 303-12, 1982.
7. H. Bauker and F. W. Haesing, "Fiber optic radiation sensors," Proc. SPIE, vol. 2425, p. 106, 1994.
8. B. D. Evans, G. H. Sigel, J. B. Langworthy and B. J. Faraday, "The Fiber Optic Dosimeter on the Navigational Technology Satellite 2," IEEE Transactions on Nuclear Science, vol. 25, no. 6, pp. 1619-24, 1978.
9. O. Ziemann, J. Krauser, P. E. Zamzow and W. Daum, "POF Handbook: Optical Short Range Transmission Systems" (2nd edition), Berlin: Springer, 2008.
10. S. O'Keeffe, C. Fitzpatrick, A. Fernandez, B. Brichard, F. Berghmans and E. Lewis, "Evaluation of PMMA optical fibres as gamma dosimeters for nuclear applications," in Proceedings of 13th International Plastic Optical Fibres Conference, Nuremberg, 2004.
11. S. O'Keeffe, A. F. Fernandez, C. Fitzpatrick, B. Brichard and E. Lewis, "PMMA optical fibres for real-time gamma dosimetry," in Proceedings of Optical Fibre Sensors Conference, Mexico, 2006.
12. S. O'Keeffe and E. Lewis, "Polymer optical fibre for in situ monitoring of gamma radiation processes," International Journal on Smart Sensing and Intelligent Systems, vol. 2, no. 3,

pp. 490-502, 2009.

13. C. Lethien, C. Loyez, J.-P. Vilcot, N. Rolland and P. A. Rolland, "Exploit the Bandwidth Capacities of the Perfluorinated Graded Index Polymer Optical Fiber for Multi-Services Distribution," Polymers, vol. 3, pp. 1006-28, 2011.

14. H. Henschel, M. Körfer, K. Wittenburg and F. Wulf, "Fiber Optic Radiation Sensing Systems for TESLA," Elektronen-Synchrotron DESY, MHF-SL Group, TESLA Report No. 2000-26, 2000.

15. GigaPOF-50SR datasheet, Chromis Fiberoptics, [Online]. Available: `http://i-fiberoptics.com/pdf/sp-fb-01_gigapof50sr.pdf`.

16. S. Liehr, M. Wendt and K. Krebber, "Distributed perfluorinated POF strain sensor using OTDR and OFDR techniques," Proc SPIE, vol. 7053, pp. 75036G-1, 2009.

17. B. Brichard, A. F. Fernandez, H. Ooms, P. Borgermans and F. Berghmans, "True doserate enhancement effect in phosphorous doped fibre optic radiation sensors," in Proceedings of 2nd European Workshop on Optical Fibre Sensors, Santander, 2004.

18. H. Zollinger, "Color Chemistry: Syntheses, Properties, and Applications of Organic Dyes and Pigments," Wiley, 2003.

19. J. J. Workman and L. Weyer, "Practical Guide to Interpretive Near-Infrared Spectroscopy" (2nd edition), Boca Raton, FL: CRC Press, 2012.

Applications of POF Sensors in the Electric Energy Sector

M.M. Werneck,[1,*] R.C.S.B. Allil,[2] C.C. Carvalho,[1] F.L. Maciel,[1] D.M. Santos,[1] F.V.B. Nazaré[1]

[1] Photonics and Instrumentation Laboratory, Electric Engineering Program, Universidade Federal do Rio de Janeiro, Cidade Universitária, RJ, Brazil.

[2] Biological Defense Laboratory, Division of Chemical, Biological and Nuclear Defense, Brazilian Army Technological Center, Rio de Janeiro, RJ, Brazil.

*Corresponding author: werneck@lif.coppe.ufrj.br

The Laboratório de Instrumentação e Fotônica - LIF (Photonics and Instrumentation Laboratory) at Universidade Federal do Rio de Janeiro is a R&D laboratory mainly involved in optical sensors applied on Energy, Oil & Gas and Biotechnology areas. This paper will show some techniques used by the LIF to measure, detect and monitor several physical parameters applying the POF either as sensors or as a communication channel for telemetry. The applications to be presented include measurement of electrical current, high voltage, temperature, leakage currents and displacement. In most cases, we take advantage of the plastic optical fibers (POF) for their high insulation property, applying them to measure electrical parameters in high voltage environment, normally found in the electrical energy area. For each example presented, it will be shown the measurement principle, laboratory tests and field application.

1. Introduction

This paper will show selected techniques used by the Photonics and Instrumentation Laboratory - LIF to measure, detect and monitor several physical parameters applying the POF either as sensors or as a communication channel for telemetry. The applications to be presented include measurement of electrical current, high voltage, temperature, leakage currents and displacement. In most cases, we take advantage of the plastic optical fibers (POF) for their high insulation property, applying them to measure electrical parameters in high voltage environment, normally found in the electrical energy area.

2. The Use of POF as Transmission Medium

Optical fiber sensors are very convenient for the electric power industry as they are dielectric and therefore can be directly applied to high voltage without security hazards or short-circuits. On top of that POF medium presents the following advantages as compared to silica fibers:

- good transmission for VIS
- high numerical aperture: Goes well with LED
- highly multimode: Low bend loss
- easy maintenance (D.I.Y. in the field)

The applications presented in this work will use a conventional configuration containing a LED, the POF and a photodetector, referred to as a LED-POF-PD assembly.

By the use of this simple configuration applied to high voltage and current measurement, one can for instance build an electronic system containing a resistive divider and a Rogowski coil to measure voltage and current in a 13.8 kV line. The system, powered by a small battery will be fluctuating in the three-phase distribution line and an optoelectronic section is used to modulate the data and transmit it through a LED-POF-PD assembly.

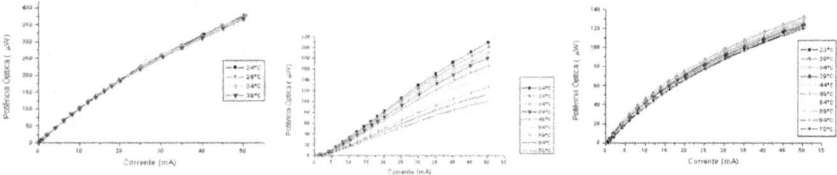

Fig. 1: Power drift of blue, yellow and green LEDs, from left to right.

In case of using the LED-POF-PD assembly for an amplitude modulated sensor, the LED as a light source, needs to be as stable as possible. The main factor to change the LED characteristics is the temperature. Therefore, one has to choose the right LED for such application. Fig. 1 shows the power drift of several LED colors against the temperature. Notice the smallest drift is presented by the blue LED.

Another important drift is the center wavelength as the Si photodetector, with its peak sensitivity on near IR, will act as an edge filter for visible wavelengths, therefore producing variable outputs when wavelength displaces. Fig. 2 shows the center wavelength for the three LEDs of Fig. 1. Again, the blue LED drifts less than others and therefore, this is our choice as a light source.

Now, considering the whole LED-POF-PD assembly, shown in Fig. 3 (left), and employing a blue LED as light source and a trans-impedance amplifier to drive the photodetector, we can

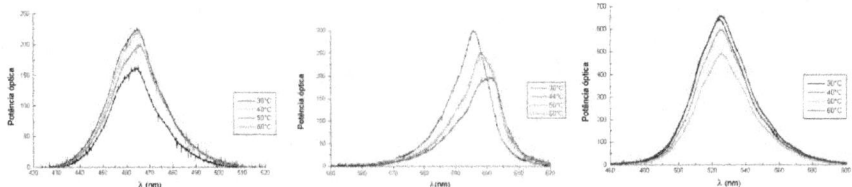

Fig. 2: Center wavelength drift of blue, yellow and green LEDs, from left to right.

monitor the voltage output against the LED input current for different temperatures. Fig. 3 (right) shows the obtained family of curves. All curves drift; however, for 5 mA input current, the output voltage drifts negligibly in the room temperature range.

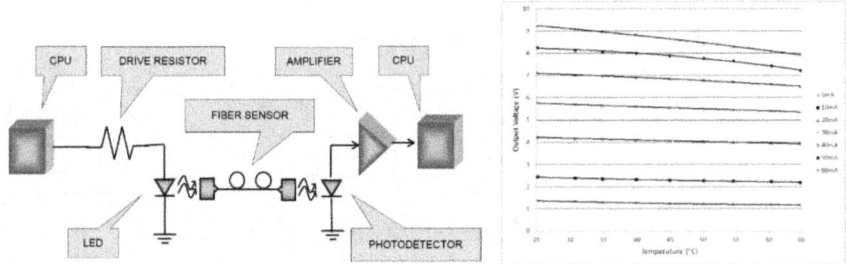

Fig. 3: (Left) Basic LED-POF-PD assembly with the photodetector driven by a transimpedance amplifier; (Right) Family of curves showing the temperature drift of the photodetector output voltage versus the LED input current.

3. Techniques for high-voltage and current sensing

By applying the LED-POF-PD assembly, several setups were developed for measuring voltage and current in high voltage environments. Fig. 4 shows the setup for measuring low current in high voltage. The current to be measured is made to cross the transmitting LED after being rectified by the four diodes in bridge. The two transorbs are used to absorb transients always present in high voltage transmission lines. The optical fiber insulator is used to increase the creepage distance between high voltage potential and ground potential avoiding leakage currents over the fiber cable surface.

Fig. 5 shows the same setup with an added resistor for measuring high voltage. The series resistor is used to control the current for the transmitting LED.

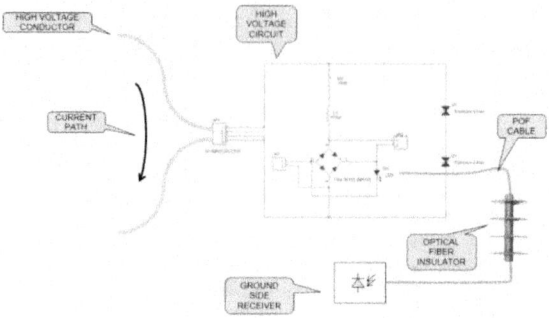

Fig. 4: Measuring low current in high voltage.

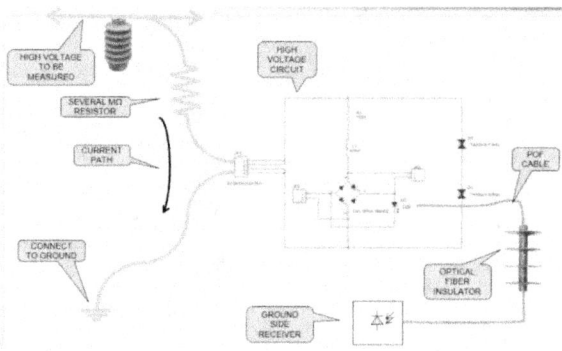

Fig. 5: Measuring high voltage.

Fig. 6 shows a slightly different setup for measuring high currents on the left. In this case, the high current is divided to an appropriate value by the current transformer (CT) shown. The CT can be of the open magnetic core type, so as to be possible to install it around the conductor without opening the circuit. Fig. 6 shows an example of transducer for measuring high current on the right. The screw on the bottom of the transducer is used for closing or opening the magnetic core around the high voltage conductor.

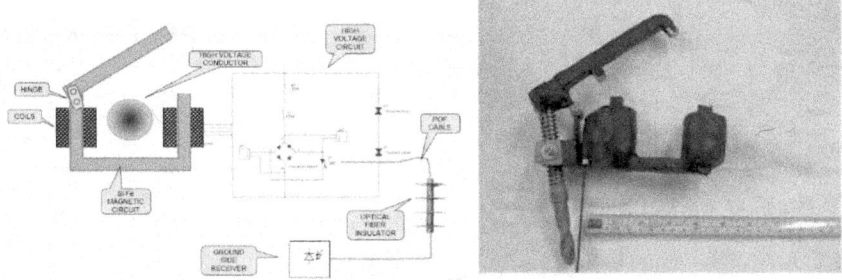

Fig. 6: (Left) Measuring high current in high voltage.; (Right). An example of high voltage current transducer.

4. Field applications

This section will describe successful field applications using the techniques shown in the last sections applied to the electric energy area.

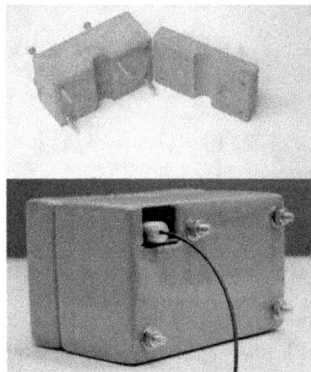

Fig. 7: Current transducer for 13.8 kV distribution lines.

4.A. Current monitoring in distribution lines

Using the setup shown in Fig. 6 it is possible to measure current in high voltage environments such as 13.8 kV distribution lines. For electric companies this information is useful for helping planning expansion of lines or repotencialization of old transmission lines. Fig. 7 shows the transducer employed. It is composed of a split-core current transformer that is mounted around the conductor. After installation, the technicians tight the assembly with screws and the transducer is left on the line during the monitoring period. Fig. 8 show an example of trace obtained. Each cycle, from 80 A to 120 A means one entire day, peak consumption around noon and minimum consumption during the night.

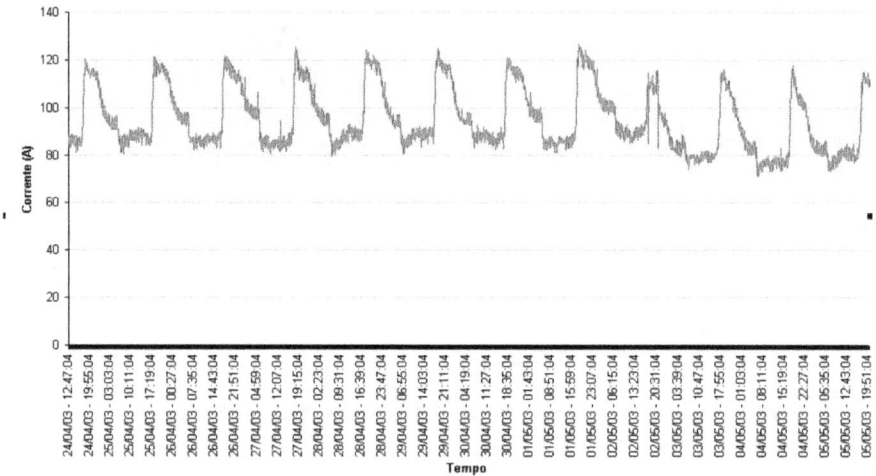

Fig. 8: Example of trace obtained with the transducer.

4.B. High-voltage leakage current

Generally, high-voltage transmission line insulators are subjected to different conductive pollution sources, which often cause an insulator's flashover and subsequent breakdown. Flashover of insulator strings is a very serious disaster affecting transmission lines. It is also dangerous to the power grid because of the potential for significant grid losses. Contamination originates from diverse substances present in the environment surrounding high power transmission lines, for example, sea salt, powder from nearby dirt roads, agricultural fires and by-products from industrial operations. These substances are deposited on the insulating surface. In dry conditions, these layers do not cause great problems, however, in wet situations, such as in the presence of rain, humidity, fog or dew, the substances start to conduct electricity from the high voltage potential to ground potential over the insulator surface; these currents are called leakage current. The pollution layer keeps increasing until it causes a failure on the high-voltage electrical transmission system. The probability of this failure occurrence depends on the nature of insulator material and type, the climatic conditions of the area, the types of pollutants, the degree of contamination, and the voltage level of the transmission line.

In order to mitigate this problem, transmission lines operators often contract a third-party service to wash the insulator with a jet of treated water. As this service has to be continu-

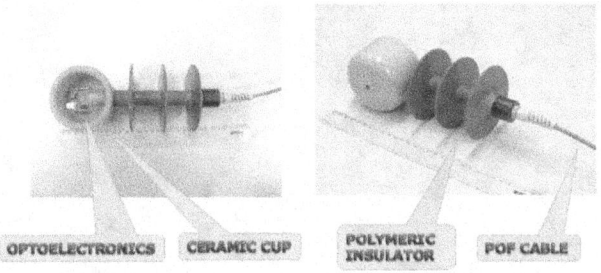

Fig. 9: Leakage current transducer.

ally performed and it is very expensive, the Photonics and Instrumentation Laboratory was contracted to develop a means to monitor the insulators pollution. We decided by a current monitoring system installed on one of the insulators. The transducer is shown in Fig. 9 and the installation setup is shown in Fig. 10. Notice that the leakage current is made to cross the transducer before flowing through the insulator surface to ground potential.

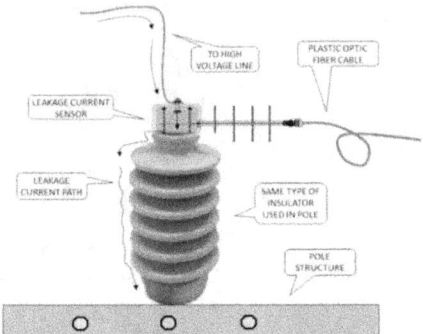

Fig. 10: Installation setup for the transducer over the insulator.

Fig. 11 shows the type of trace obtained with such setup in a 13.8 kV distribution line. In the graph, it is possible to see the leakage current, the humidity, the local temperature and the dew point. When the temperature drops during the night and reaches the dew point, condensation occurs on the insulator surface. This melts the salt deposited on the insulator, becoming conductive. Notice that the leakage current increases just when the dew point is reached.

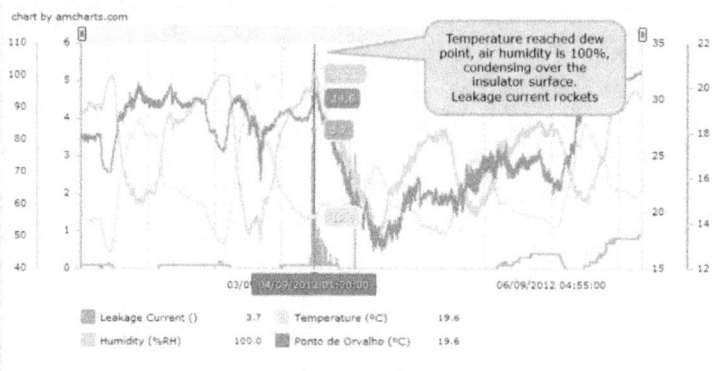

Fig. 11: Results for a 13.8 kV transmission line.

The same transducer can be used for monitoring the leakage current on high voltage transmission lines, such as 500 kV. Fig. 12 shows the installation setup on the tower steel structure. A copper conductor is used to bypass the leakage current from the last insulator of the string to the sensor and them to ground potential.

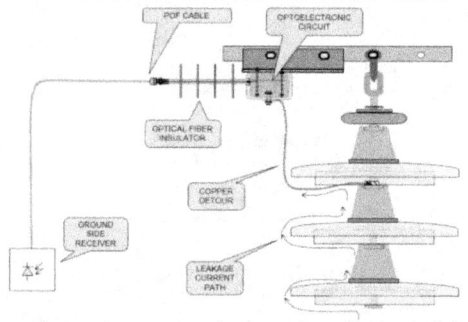

Fig. 12: Installation setup for a 500 kV transmission line.

Fig. 13 shows the obtained results for a 500 kV transmission line. The upper trace is the temperature that goes from 25°C during the night to 40°C at noon. The leakage current in turn, is nearly zero during the day but increases during the night as temperature drops eventually reaching the dew point. Note also, that there is a leakage current growing in a day to day basis after the occurrence of a rain that washed the insulator.

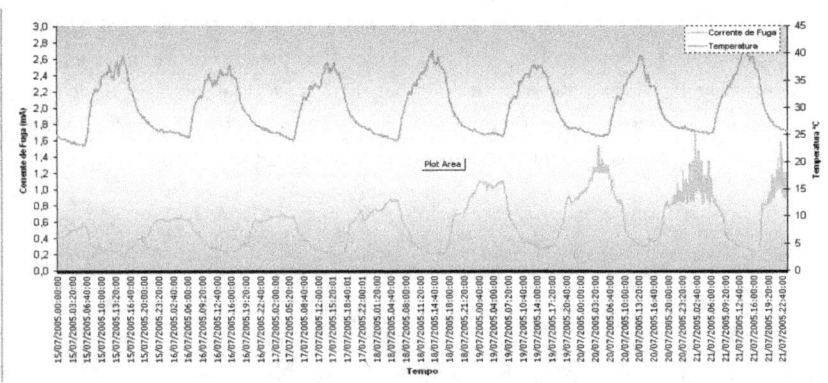

Fig. 13: Leakage current for a 500 kV transmission line.

4.C. Temperature monitoring of high-voltage harmonic filters

HVDC transmission technique firstly rectifies the AC power signal and transmits it in a two wire transmission line. At the load side, the DC has to be converted back to AC by a combination of thyristors. The resulting output sinusoidal wave shape is similar to the one shown in Fig. 14 (left). It is a staircase-like sinusoidal signal that contains odd harmonics, particularly the third and the fifth, as shown on the right.

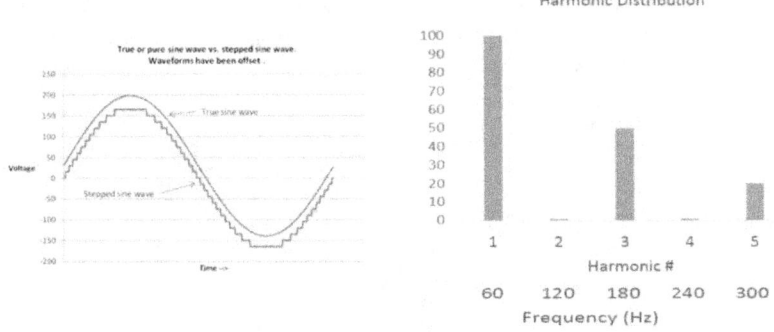

Fig. 14: (Left) Staircase type wave form after AC recovery; (Right) Harmonic distribution.

Brazilian Itaipu power plant at southwest of Brazil, border with Paraguay, transmits its 14 GW of power over two 780 km 600 kV HVDC lines. At the load side, in the State of São Paulo, after the converter building there is a large filter to eliminate the third and fifth harmonics. Fig. 15 shows a picture of the filter where the inductors are indicated by arrows. The filter impedance is shown in Fig. 16 (left). Notice that the impedance modulus is zero with phase zero at 180 Hz and 300 Hz, meaning that there is a virtual short-circuit do ground at these frequencies. Due to dissipation power and high currents, the inductors of this filter tend to heat up and that is the point where an optic fiber sensor has its role.

Fig. 15: The AC filter with the inductors indicated by arrows.

The Photonics and Instrumentation Laboratory was contracted to measure the temperature of the filter coils. The solution we found was to build a temperature sensor based on the fluorescence of ruby. The gemstone sapphire contains pure alumina (Al_2O_3) whilst the gemstone ruby contains the same aluminum oxide of sapphire but with chromium: Al_2O_3:Cr. Artificial ruby is used in mechanical watches to reduce friction, increasing accuracy and bearings life. Artificial ruby, when excited with blue light, produces red fluorescence, as shown in Fig. 16 (right).

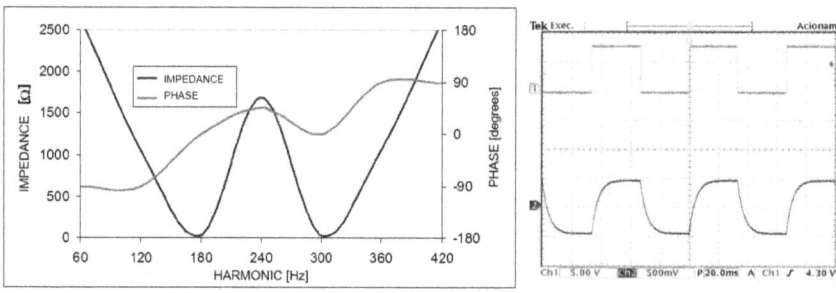

Fig. 16: (Left) Modulus and phase of the AC filter; (Right) Pump light and resulted fluorescence of an artificial ruby.

Notice in Fig. 16 on the right that the fluorescence light presents a decay time. Fig. 17 shows on the left the decay time for different temperatures in an exponential fashion. The decay time can be plotted against temperature, producing the graph shown in Fig. 17 on the right. Note the linear relationship between decay time and temperature with a sensitivity of 22.5 μs/°C.

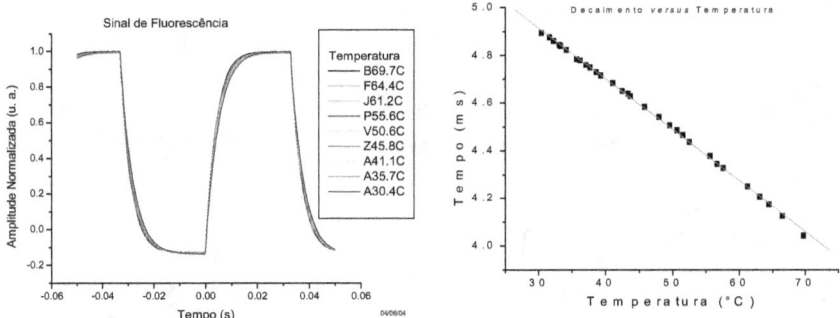

Fig. 17: (Left) Decay time of ruby for different temperatures; (Right) Linear relationship between decay time and temperature.

The sensor is based in an artificial ruby bonded to the tip of a 1 mm diameter POF. At the other extremity of the POF, a blue LED pulses the light guided to the ruby, which produces red light. The fluorescence light returns by the same fiber reaching a photodetector after the 3-dB divider. The setup is shown on the left in Fig. 18, whilst the right shows the spectra. The blue line is the LED spectrum whilst the red line is the fluorescent spectrum. The green line is the transmittance of the red filter, used to block some of the blue light that also returns to the photodetector.

The optoelectronic setup is shown in Fig. 19 on the left. The computer calculates the decay time of ruby by an exponential interpolation. The probe is shown on the right. Four probes were installed in the filter coil, two on the top and two on the bottom.

Fig. 20 shows the results. On the left it is possible to see the four temperatures plus the ambient temperature. Note that the maximum temperature does not exceed 70°C. On the right one can see a shortage occurred by 1:00 AM, in which the four temperatures reach the ambient temperature (red line). Then, by 9:30 AM the energy is on again and the four temperatures rise to normal values.

4.D. High-voltage switch monitoring

A high-voltage switch is just like a wall switch but for high voltage and high currents. It is used to connecting and disconnecting of transmission lines or other components to and from

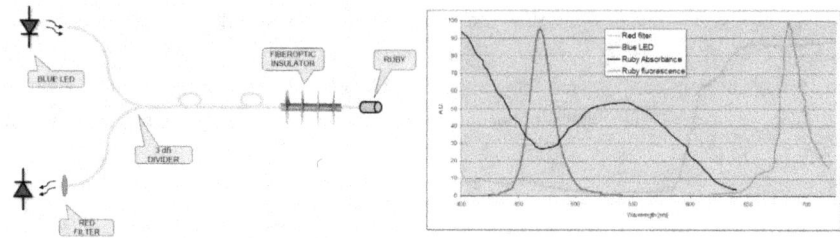

Fig. 18: (Left) The optical setup; (Right) LED spectrum (blue line); Fluorescence of ruby (red line); Transmittance of the red filter (green line) and Absorbance of ruby (black line).

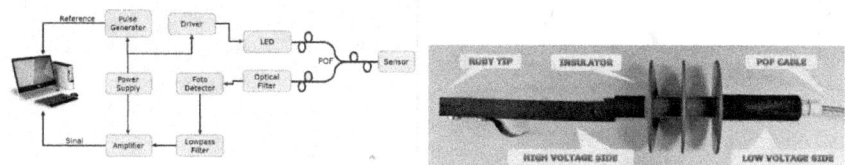

Fig. 19: (Left) The optoelectronic setup; (Right) The probe.

the system, disconnecting an entire city or an equipment into a substation for maintenance. It is composed of a long aluminum arm that closes or open the circuit between two insulated contacts, each one on the top of a ceramic insulator. The contact resistance must be in the order of a few meters or less, in order to dissipate the minimum power. For a current of 8,000 A and a resistance of 10 mΩ, we would have a dissipation of 1 W! ($P = IR^2 = 8,000\,\text{A} \times (10\text{m}\Omega)^2 = 1\,\text{W}$).

If the contact is not well formed a high power dissipation will occur and the aluminum contact may melt, destroying the switch. However, there is no established means to know whether a contact is well closed and the solution to circumvent this lack of information is to send a worker to the base of the switch to look at the contact and radio the information to the control room.

In order to propose a sensor to indicate the perfect closing of the three contacts, the Photonics and Instrumentation Laboratory was contracted. The proposed method was the measurement of the movement of the contact blade. Every time the arm locks into the contact head, the contact blades move a few millimeters. Therefore the sensor has to be a high precision displacement sensor. This was performed using a reflective power amplitude system in which a mirror, driven by the contact blade, approaches or moves away from the POF tip. Fig. 21 shows the idea on the left and a block diagram of the optical setup on the

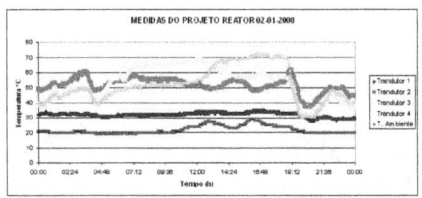 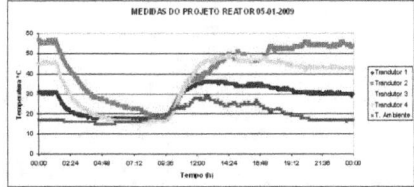

Fig. 20: The four temperatures and the ambient temperature (red line). On the right one can notice a shortage occurred by 1 AM, in which the four temperatures reach the ambient temperature (red line). Then, by 9:30 AM the energy is on again and the four temperatures rise to normal values.

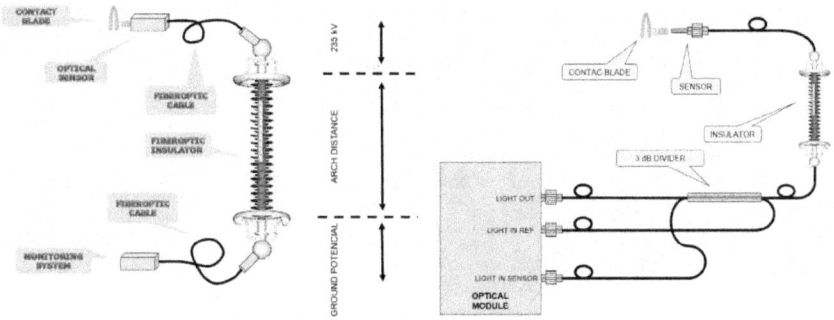

Fig. 21: (Left) The sensing system general idea; (Right) The optical setup.

right. The sensor is a 1-mm SMA connector proper for a 1-mm PMMA POF.

Fig. 22 shows the developed device, completely water proofed and adequate to the harsh environment found in outdoor power substations. The total displacement of the contact blade is about 300 µm and the sensor has to have a sensitivity in the order of 10 µm. Fig. 23 shows the output power versus blade displacement.

Fig. 24 shows the prototype installed on a real contact head aon the left and a picture of the switch in closed state on the right, with the movable arm fit inside the contact head.

After laboratory tests, three sensors were installed in a real switch, in Brazil's southwest Coxipó substation. Fig. 25 shows a picture of the installation stage on the left. Finally, three industrial prototypes were developed by a spinoff company and are to be installed in an operating switch for commissioning tests.

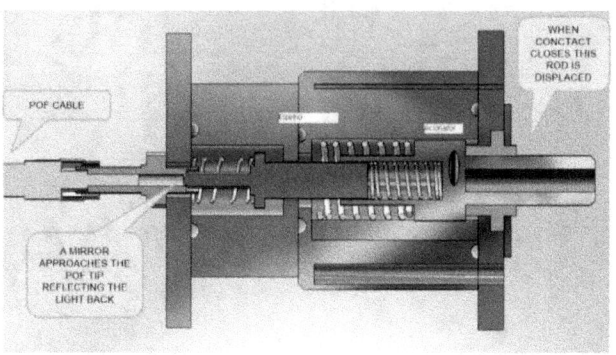

Fig. 22: The transducer.

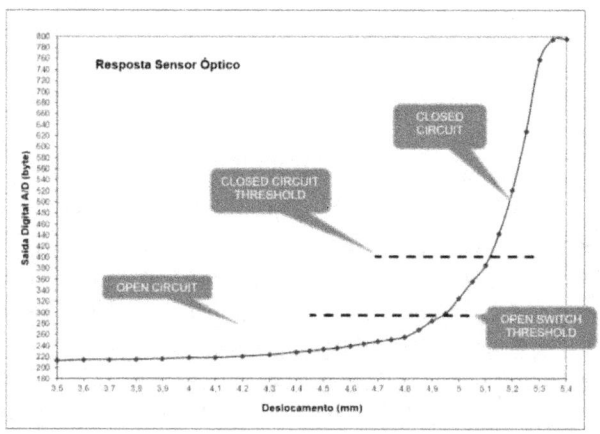

Fig. 23: Graph showing the output power versus blade displacement.

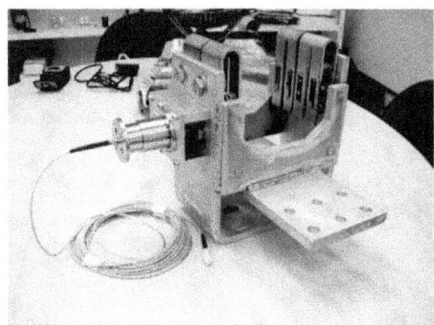 

Fig. 24: (Left) The prototype installed on a real contact head; (Right) Picture of the switch in closed state, with the movable arm fit inside the contact head.

 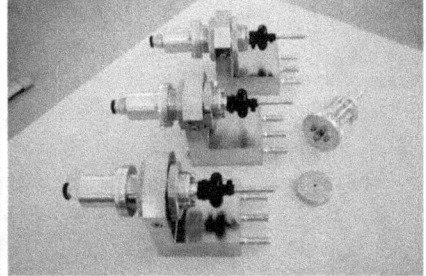

Fig. 25: (Left) The installation of a prototype in a real switch; (Right) Three industrial prototypes of the transducer.

5. Conclusion

This paper demonstrated applications of POF sensors applying an amplitude modulated approach. All sensors were developed in laboratory and further applied in the field in real applications.

We presented current sensors for both mA and kA, voltage sensor, displacement sensor and temperature sensor. The resultant transducers from these researches proved to be reliable, easy to build and calibrate and of low cost.

Acknowledgements

The authors acknowledge the support of the Brazilian research foundations such as CNPq, FAPERJ and FINEP.

We also acknowledge the support of electric energy companies as sponsors of some projects, such as Eletronorte, Ampla and Furnas.

Finally, we acknowledge the support of the students and technicians of the Laboratório de Instrumentação e Fotônica. I thank you all, acknowledging that these projects would not succeed without your cooperation and support.

References

1. Marcelo M. Werneck, Sydney B. Germano, Ricardo M. Ribeiro, Fernando L. Maciel, Plínio Porciúncula, André Almeida and Luciana Martins, "Plastic Optical Fiber Technology for High Voltage Leakage Current Monitoring". Proceedings of the 11th International POF Conference 2002, pp 271-273, Hotel New Otani, Tokyo, Japan, September 18-20, 2002.
2. Marcelo M. Werneck, Sydney B. Germano, Ricardo M. Ribeiro, Fernando L. Maciel, Plínio Porciúncula, André Almeida and Luciana Martins, "Plastic Optical Fibre Technology for High Voltage Current Measurements". Proceedings of the International POF Conference 2002, pp 275-278, Hotel New Otani, Tokyo, Japan, September 18-20, 2002.
3. Marcelo M. Werneck, Fernando L. Maciel, Cesar C. Carvalho, Ricardo M. Ribeiro, "Development and field tests of a 13.8 kV leakage current LED/POF based sensor". Proceedings of the 12th International POF Conference 2003, pp 54-57, University of Washington, Seattle, EUA, September 14-17, 2003.
4. Marcelo M. Werneck, Cesar C. Carvalho, Ricardo M. Ribeiro and Fernando L. Maciel, "High-voltage current sensing based hybrid technology". Proceedings of the 12th International POF Conference 2003, pp 50-53, University of Washington, Seattle, EUA, September 14-17, 2003.
5. Ricardo M. Ribeiro, L.A. Marques-Filho and Marcelo M. Werneck, "Fluorescent plastic optical fibers for temperature monitoring". Proceedings of the 12th International POF Conference 2003, pp 282-285, University of Washington, Seattle, EUA, September 14-17, 2003.

6. Marcelo M. Werneck and L. R. Kawase, "POF Activities in Brazil: A Brand new Market". Proceedings of the 12th International POF Conference 2003, pp 259-262, University of Washington, Seattle, EUA, September 14-17, 2003
7. Marcelo M. Werneck, Cesar C. de Carvalho, Ricardo M. Ribeiro e Fernando L. Maciel "Desenvolvimento de Sistema de Monitoramento de Corrente para Classe de Tensão de 13,8 a 138 kV". Anais do II Congresso Nacional de Inovação Tecnológica do Setor de Energia Elétrica, Salvador, Brasil, Novembro, 2003.
8. M.M. Werneck, C.C. Carvalho, R.M. Ribeiro, F.L. Maciel, "Application of a POF-based current sensor for measuring leakage current in 500 kV transmission line". Proceedings of the 13th International Conference on Polymer Optical Fibre - ICPOF2004, pp 345-350, Nürnberg, Germany, from 27 to 30 September 2004.
9. R.M. Ribeiro, M.M. Werneck and L Marques-filho, "Simple and low cost temperature sensor using the ruby fluorescence and plastic optical fibres," Proceedings of the 14th International Conference on Polymer Optical Fibre - ICPOF2005, pp 291-294, held in the Sheraton Hong Kong Hotel and Towers, Hong Kong, from 20th September to 22nd September 2005.
10. M. M. Werneck, "POF sensors and systems at photonics and instrumentation laboratory", Proceedings of the 15th International Conference on Polymer Optical Fibre - ICPOF2006, "The Joint International Conference on Plastic Optical Fiber & Microoptics 2006", held in the Grand Hilton Seoul, Seoul, South Korea, pp 265-271, 11th to 14th September, 2006.
11. M. M. Werneck, E. S. Yugue, F. L. Maciel, C. C. Carvalho, A. V. da Silva, J. L. Silva Neto, M. A. L. Miguel, R. M. Ribeiro, "Application of a POF and ruby florescence based temperature system in an electric power substation." Proceedings of the 16th International Conference on Plastic Optical Fibers - POF 2007, Turin, Italy, pp 25-28, 10th to 12th September, 2007.
12. M.M. Werneck, J.Zubia, H. Poisel, D. Kalymnios, K. Krebber, P.Sully. "POF Sensors applicalions in every day's life", 33rd European Conference and Exhibition on Optical Communication, (ECOC 2007) – International Congress Center (ICC), Berlin, Germany, September 16th to 20th, 2007.
13. M. M. Werneck, "POF Sensors and Systems at The Photonics and Instrumentation Laboratory" IV Internacional Symposium on Non – Crystalline Solids, VII Brazilian Symposium on Glass and Related Materials 4th Internacional School on Glasses and Related Materials, Aracaju – Sergipe, Brasil, 21 a 28 de Outubro de 2007.
14. M. M Werneck, A.V. Silva, C.C. Carvalho, N.C.C Souza, M.A.L Miguel, C. Beres, E.S Yugue, F.L.Maciel, J. S, Neto e C. R.F Guimarães, R.C.B Allil, J. A. G. Baliosian ."Fiberoptic applications in sensors and telemetry for the electric power industry". 1st Workshop on Specialty Optical Fiber and their Applications - 1st WSOF 2008,Volume 1055, pp 43-45, São Pedro - SP, August 20-22, 2008.
15. M. M. Werneck; C. C. Carvalho, D. M. Santos, J. L. da Silva-Neto, F. L. Maciel and

R. C. Allil, "Development and field tests of na LED/POF-based leakage current sensor industrial prototype for 13.8 kV distribution lines", Proceedings of the 21st International Conference on Plastic Optical Fibers - POF 2012, Atlanta, USA, 10th to 12th September, 2012.

16. Marcelo Martins Werneck, Daniel M. Santos, Fábio V. B. de Nazaré1, J. L. da Silva Neto, R. C. Allil, B. A. Ribeiro, C. C. Carvalho and F. Lancelotti, "Detection and Monitoring of Leakage Currents in Distribution Line Insulators," Proceedings of the IEEE International Instrumentation and Measurement Technology Conference (I2MTC 2014), pp 468-472, Radisson Montevideo Victoria Plaza Hotel & Conference Center, Montevideo, Uruguay, May 12-15, 2014.

17. D. M. Moreira, F. Lancelotti, F.V.B. de Nazare, C.C. Carvalho, M.M. Werneck, "Monitoramento de Corrente de Fuga em Isoladores de Distribuição Utilizando Fibra Óptica Plástica", XXI Seminário Nacional de Distribuição de Energia Eletrica – SENDI-2014, Mendes Convention Center, Santos, SP, 10 a 13 de Setembro, 2014.

18. Daniel Moreira dos Santos, Cesar C. Carvalho, Marcelo M. Werneck, "Avaliação do Grau de Degradação de Pararraios Através da Análise do THD de sua Corrente de Fuga", apresentado no XVI ERIAC - ENCUENTRO REGIONAL IBEROAMERICANO DE CIGRÉ (La Bienal de Cigré en Iberoamérica), Puerto Iguazú, Argentina, de 17 a 21 Maio de 2015.

19. Marcelo Werneck, "Biosensors with POF Technology", invited talk at the 24th International Conference on Plastic Optical Fibers, held at Technische Hochschule Nürnberg Georg Simon Ohm, Nuremberg, Germany, 22nd to 24th September, 2015.

20. Marcelo Martins Werneck, "APPLICATION OF POF SENSORS IN ENERGY, OIL AND BIOTECHNOLOGY", 3rd International POF Modelling Workshop 2015, September 21, 2015, Nuremberg, Germany, joint with the 24th International Conference on Plastic Optical Fibers, held at Technische Hochschule Nürnberg Georg Simon Ohm, Nuremberg, Germany, 22nd to 24th September, 2015.

21. Werneck M, Santos D, Carvalho C, de Nazaré F, Allil R, "Detection and Monitoring of Leakage Currents in Power Transmission Insulators" DOI 10.1109/JSEN.2014.2361788, IEEE Sensors Journal, ISSN: 1558-1748, September 2014.

The Influence of Equilibrium Mode Distribution on an Optical Power Monitor for SI PMMA Polymer Optical Fibre Based on Side-Scattered Light

T.A.M.G. Freitas,[1,*] R.M. Ribeiro,[1] V.H.N. Silva,[1] C.B. Marcondes,[1]
P.A.M. dos Santos,[2] J. Fischer,[3] O. Ziemann,[3] H. Poisel,[3] R. Kruglov[3]

[1] *Laboratório de Comunicações Ópticas (LaCOp), Departamento de Engenharia de Telecomunicações – Universidade Federal Fluminense – 24.210-240, Niterói, RJ – Brasil.*

[2] *Laboratório de Óptica Não-Linear e Aplicada, Instituto de Física - Universidade Federal Fluminense - 24.210-346, Niterói, RJ – Brasil.*

[3] *Polymer Optical Fiber Application Center (POFAC) - Technische Hochschule Nürnberg Georg Simon Ohm Wassertorstrasse 10, 90489, Nuremberg – Germany.*

Corresponding author: taiane@telecom.uff.br

We show the influence of the mode distribution on the performance of an optical power monitor (OPMo). The latter is based on the spontaneous side-scattered (visible) light intended for SI PMMA polymer optical fibre links. At first, it was shown by using a mode-scrambler that such OPMo is highly immune to the modal distribution of light. At second, measurements using a scanning integrating sphere have shown that a couple of meters length of POF after the light source is enough to significantly attenuating the higher order and cladding modes. As a result, the modes guided by the core even before to reach the equilibrium mode distribution (EMD) condition, does not appreciable affect the side-scattered light thus rendering the OPMo stable. A theoretical approach based on the power flow equation is suggested.

1. Introduction

Optical power monitors (OPMos) are very useful and well-known active devices intended to monitor/measure the optical power flowing along optical fibers. It can be temporarily or permanently deployed inline [1,2] as is schematically shown in Fig. 1.

Fig. 1 sketch a general optical fibre link where an OPMo is inline inserted thus monitoring and measuring the propagating optical power. It does not require the interruption of the data stream, except in the deployment. Therefore, OPMos require a smaller insertion loss as much

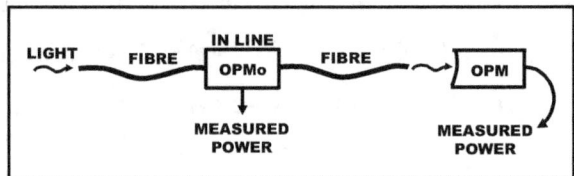

Fig. 1: The schematic drawing of a fibre-optic link using an (inline) OPMo and an Optical Power Meter (OPM) in the end of the line.

as possible. For comparison purposes, a conventional OPM is depicted in the end of the line. OPMs are also able to measure optical power, but require the interruption of the data stream.

OPMos are useful devices for real-time in single or multiple-points monitoring. In this way, OPMos can be used for monitoring of parameters on optical devices as the attenuation of Variable Optical Attenuators (VOAs), gain of optical amplifiers, etc. OPMos are also used for monitoring of optical fibre links in general as simple point-to-point links, branches of networks, fibre-optic sensors and fibre circuits under development in laboratories.

Most of commercially available OPMos uses a 1-10% tap-coupler to derive an optical power sample to be measured [3]. Fig. 2 shows the typical configuration of a tap-coupler based OPMo.

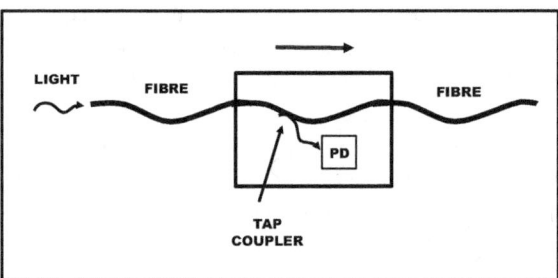

Fig. 2: The schematic drawing of an OPMo using a tap-coupler to derive an optical power sample.

As can be seen in Fig. 2, an OPMo based on tap coupling is able to measure the optical power flowing in a single direction as assigned by an arrow. For bidirectional measurement capability, an additional tap-coupler and the respective photo-diode are needed to collect and measure, respectively, the optical power flowing in the opposite way. The tap-coupler based technique allows stable measurement for single-mode fibers. However, tap-couplers, couplers and then OPMos for multimode fibers based on tap coupling require the EMD condition to result in reproducible measurements [4,5].

The margin of power budget for POF links is generally narrow. Therefore, connections should be avoided and when they are needed it should be carefully performed [6]. Therefore, this is the main reason that OPMos for POF links should present a minimum insertion loss. Furthermore, regarding the characteristics of POF technology [6], it is easy to conclude that a suitable OPMo should be also reliable, simple and as much inexpensive as possible.

We have been developing OPMos for PMMA SI-POF links operating in the visible spectrum [1,2,7,8]. The OPMo is non-invasive because it collects and detects the spontaneous and unavoidable Rayleigh and Mie side-scattered light [9]. It does not tap any guided light from the fibre core using tap-coupling, fibre-bending or evanescent coupling [2]. In other words, the OPMo operates without causing any disturbance in the light propagation. The OPMo works bi-directionally and indicates whether a POF transmission link is in the *dark* or in the *live* status. In the latter, it measures the average optical power level of the propagating signals without disconnecting the fibre link. It was achieved 45-dB dynamic range and − 50-dBm sensitivity [2], a very promising milestones for a further OPMo development toward to the marketplace.

The present paper describes our further experimental findings regarding the OPMo [2] in which the amount of collected spontaneous Rayleigh and Mie [9] side-scattered light seems to be *independent* of the mode distribution.

2. The OPMo with mode-scrambler

2.A. Experimental

Fig. 3 is a schematic drawing of the OPMo and the experimental set-up [2] used to their characterization. A homemade JIS6863 model mode-scrambler shown in the bottom left inset of Fig. 3 was connected just before the OPMo. The mode-scrambler is built from 3.4 m POF length that was wrapped in figure-of-eight around two metallic cylinders. A 5mm ultra-bright green LED with transparent plastic dome emitting centered on 525 nm wavelength is used as the light source probe. The optical coupling into the standard POF was carried out by using the butt-to-butt technique. As is shown in the top right inset of Fig. 3, the photo Darlington module was extracted from the plastic housing of an IFD93 model of Industrial Fiber Optics (USA). The 1.5-mm diameter micro-lens of the photo-Darlington placed < 1 mm apart from the POF surface was able to collect the light radiated from 1.5 mm of stripped POF length (11 mm in total) and it is focused onto the photo-Darlington chip. A picture of the stripped POF is shown in the bottom right inset of Fig. 3.

A voltmeter was used to probe the limiting-current 68-kΩ resistor of the photo-Darlington circuit that was electrically polarized with 5 V DC, typically measuring few sub-mV and mV voltage electric signals. The output light from the OPMo follows through the same PMMA POF. The optical power was externally measured by means of a high-performance calibrated

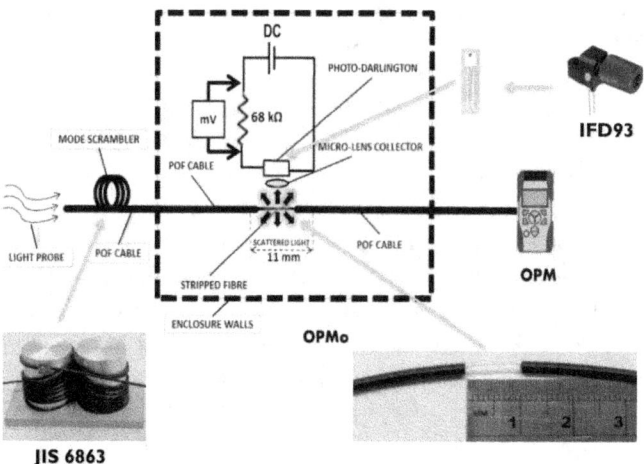

Fig. 3: The schematic drawing of the OPMo surrounded by a dashed line (enclosure walls) and the experimental set-up for their characterization. The OPMo uses a micro-lens to collecting the scattered light. Insets: Photography of the JIS6863 mode-scrambler where 3.4 m POF length is wrapped in figure-of-eight around two metallic cylinders and the stripped POF.

optical power meter (OPM) that uses a silicon photo-detector head. The conventional hand-held PM20 model from Thorlabs (USA) presenting a sensitivity of -60 dBm was used. All measurements were carried out in real time, that is, without employing any averaging procedure to increase the signal-to-noise ratio.

2.B. *Results and discussion*

Figure 4 shows the measurement results of the OPMo response, i. e. the electrical output (in dBv) against the optical power input (in dBm), that were carried out with (black square) and without (red circle) the use of the mode-scrambler (MS). The use of an MS just before the OPMo guarantees the EMD regime. More precisely, three pairs of measurements (with and without the MS) were carried out in three different days in order to check the reproducibility of the results.

In order to perform the "without-MS" measurements, the fully 3.4 m POF length originally wrapped in the MS was carefully uncoiled and extended but keeping the OPMo mechanically stable. As can be seen in Fig. 4, essentially the same plots were achieved for the three measurement sets thus showing their reproducibility. It is also clear from the plots of Fig. 4 that the OPMo displays the same response when the light probe is under EMD or not.

These experimental results suggest that our OPMo for SI PMMA POFs seems to be highly

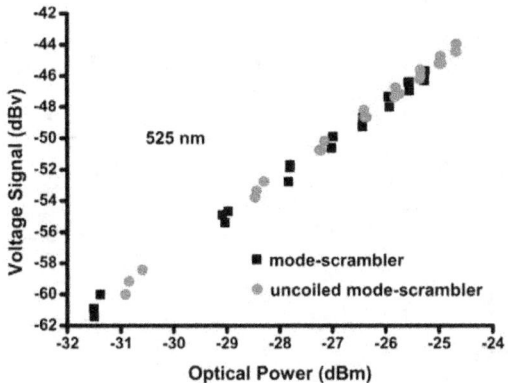

Fig. 4: Response of the OPMo without and with the use of the JIS6863 mode-scrambler. All data points follow the same straight-line dependence in a log-log plot.

immune to the mode distribution. In other words, despite the use of the JIS6863 mode-scrambler, the response of OPMo is the same. This is in contrast with the conventional OPMos for multimode fibre based on the use of tap-coupler where the coupled fraction is dependent on the mode distribution.

3. Characterization of side-scattered radiation from the SI PMMA POF

In order to better understand the previous results, the integral optical power of the side-emitting radiation is measured as a function of the launched numerical aperture (NA). The latter may be regarded as a physical simulation of an input mode distribution.

3.A. Experimental

The side-emitting radiation was measured by using an integrating sphere that can be displaced along the bare fibre for the different excitation conditions [10]. Figure 5 sketches the experimental set-up to axially measuring the lateral irradiance from the POF as a function of the numerical aperture of the launched light beam.

At first, the characterized POF was the PFU-FB1000 model without polyethylene coating from Toray Industries (Japan) which features $NA = 0.46$ and 150 dB/km at 650 nm attenuation coefficient. At second, a different SI-POF now the DB-1000 (A) model from Asahi Kasei also without polyethylene coating was probed at 532 nm.

3.B. Results and discussion

Side-emitting POFs are based on the spontaneous or enhanced light scattering and have been investigated for lighting and decoration purposes in many applications [11]. The radiation

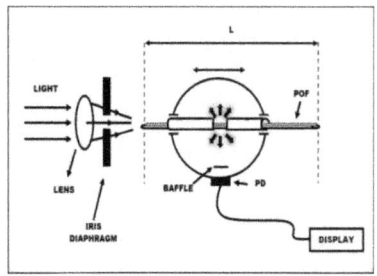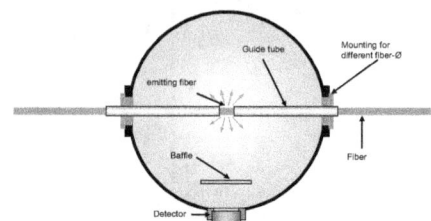

Fig. 5: Experimental set-up of the side-scattering measurements using scanning with an integrating sphere where $L = 2.5$ m.

launched that propagates along the POF can be separated in three different components: the side-emitted light due scattering, the guided light (including the fraction scattered inside the acceptance cone) and the absorbed light. The major portion of the scattered light is emitted through the side surface and a minor fraction is forward and backward coupled in the acceptance angle of the fibre.

Figure 6 shows three plots of the integral optical power of the side-emitted radiation measured from the 5°, 30° and 60° angles of launched light beam into the PFU-FB1000 SI-POF, i. e. for different numerical apertures inputs. The plots were each one normalized with the corresponding optical power launched into the POF.

The plot in red (60°) shows that a relatively high radiation power fraction is side-scattered in the first 200 mm POF length. Because the 60o input angle is wide, a large number of higher order modes are thus excited including the cladding modes. Such modes are typically attenuated or radiated out of the POF at 50-dB/m rate. From 400mm length the radiated output power for either launched NA is practically clamped to a constant level because most of the higher order modes was filtered out. For input at 30° angle a similar behavior is observed when compared with the 60° in the $z \leq 200$ mm range, but with smaller output irradiance. For inputs at $< 30°$ launching angles, it can be seen that the irradiance is not significantly different in the z 500mm range because by launching smaller angles, lower order modes are rather generated. The fibre scanning was conducted up to 2.3 m length and the plots suggest a "slow" convergence to a single level of radiated power. However, for launching angles smaller than 30° and from 700 mm it can be observed a high immunity to the launched NA.

From 0 to 500 mm POF length, there is much coupled light in the cladding that it is very rapidly attenuated by propagation. Therefore, a reduced sensitivity to launched NA is observed for $L > 0.5$ m. As a result, the light power turn to be concentrated in the core, but the EMD is not reached yet. It was already experimentally demonstrated that the EMD for SI PMMA POFs is usually reached after a length in the 8-100 m range of propagation [12] depending on the POF characteristics. In most situations around 60-70 m length is required to reach

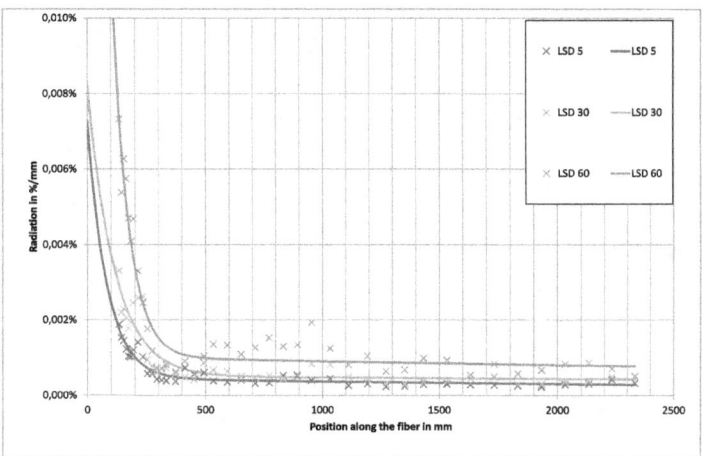

Fig. 6: Plots of the side-scattered integral radiation as a function of excitation (numerical aperture) conditions for the Toray Industries PFU-FB1000 SI-POF.

the EMD. Nevertheless, the OPMo response is not sensible to the EMD condition [2]. These results could be useful for the better understanding of the OPMo operation under EMD and non-EMD conditions.

Figure 7 shows four plots of the integral optical power of the side-emitted radiation measured from the collimated, $NA = 0.044$, 0.26 and 0.5 launched light beam into the DB-1000 (A) SI-POF. Again, the plots were each one normalized with the corresponding optical power launched into the POF.

The plot in red ($NA = 0.5$ or $60°$ launching angle) similarly shows in comparison with Fig. 6 that a relatively high radiation power fraction is side-scattered in the first 300 mm POF length. Again, a large number of higher order modes are thus excited including the cladding modes that are highly attenuated as z increases. From 500 mm length the radiated output power for either launched $NA \leq 0.26$ is practically clamped to a constant level because lower order modes are rather generated and most of the higher order modes was already filtered out. It must be observed that the corresponding plot for $NA = 0.5$ presents side-emitted power significantly higher than the either of the plots achieved for $NA \leq 0.26$. However, all the four plots clearly converge to same level for L slightly higher than 2.5 m. The later is the typical required POF length needed to reach a reduced sensitivity or even immunity to the launched NA.

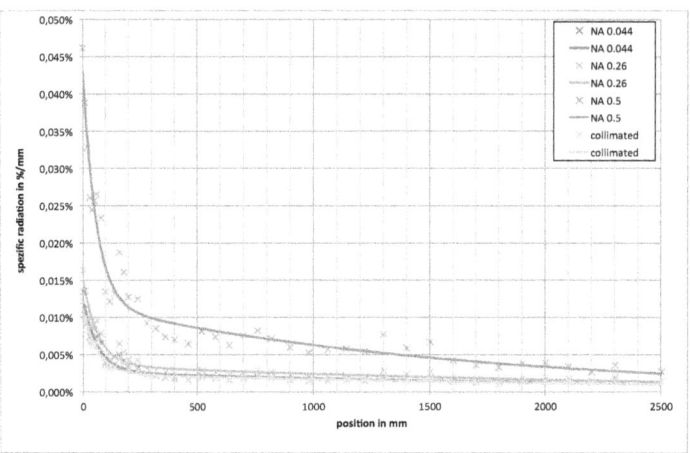

Fig. 7: Plots of the side-scattered integral radiation as a function of excitation (numerical aperture) conditions for the Asahi Kasei DB-1000 (A) SI-POF at 532 nm.

3.C. *Preliminary theoretical analysis*

The diffusion model of Gloge [6,13] leading to the power flow equation is suggested as the theoretical modeling approach.

$$\frac{\mathrm{d}P}{\mathrm{d}z} = -A\theta^2 P + \frac{D}{\theta}\frac{\partial}{\partial \theta}\left(\theta \frac{\partial P}{\partial \theta}\right) \qquad (1)$$

The power flow equation must be solved under the assumption of lack of cladding modes, in the non-EMD and EMD conditions. A typical SI MM POF can accept the propagation of a very large number of modes (millions) with deviation to the Gaussian shaped transversal distribution when the modes are out of the EMD regime. However, since the higher order and cladding modes are suppressed, the light is relatively concentrated in the core. The EMD is reached after 60-70 m length propagation and the distribution turn to be Gaussian. The experimental results have shown that the integral normalised side-scattered optical power from the fibre is highly immune to the detailed transversal distribution of the light, since the higher order meridional rays, skew rays and cladding modes are all suppressed.

4. Summary and conclusions

In a first step of experiments, our OPMo did not shown different performance when the JIS6863 was used or when their wrapped POF was uncoiled. In other words, the OPMo has shown essentially the same response when the entrance light is in the EMD or non-EMD regime.

In a second step of experiments, the integral side-scattered irradiance was measured using the axial scanning technique with an integrating sphere. This second step has pointed out that it is not allowed that the OPMo stay inline deployed at only a couple of meters lengths separated from the optical transmitter. Of course the "threshold" may vary for different POFs. Because it was used 3.4 m ($>$ 2.5 m) of uncoiled or wrapped POF length between the LED and the OPMo during the first round of experiments, the insensitivity to modal distribution was indeed expected. However, after few centimeters or meters of POF length, it is not to expect that the EMD to be reached, but an almost complete missing of meridional high-order modes, skew rays, and cladding modes. Therefore, the OPMo based on side-scattered light, requires only the higher order and cladding modes suppression and not necessarily the EMD condition!

In conclusion, our OPMo did not provide reliable performance when the light propagates in the vicinity or in the cladding, but it has shown insensitivity to the mode distribution in the fibre core, thus being a technological plus advantage when compared to the conventional OPMos. A theoretical model based on the power flow equation is suggested to support the experimental data presented in this paper and remains to be developed.

Acknowledgements

The authors would like to thank Dr. Randy Dahl of Industrial Fiber Optics (USA) for the fruitful discussions and the DFG/Germany and Brazillian R&D agencies Faperj, CNPq and Capes for the financial support.

References

1. R. M. Ribeiro, T.A.M.G. Freitas, A.P.L. Barbero, P.S.T.C. Cyrillo, W.S. Zanco and O.S. Xavier, "A Novel Optical Power Monitor (OPM) for Plastic Optical Fibre (POF) Links", 20th International Conference on Plastic Optical Fibers (ICPOF 2011), Bilbao, Spain, September 14-16, 2011.
2. Ricardo M. Ribeiro, Taiane A.M.G. Freitas, Andrés P.L. Barbero and Vinicius N.H. Silva, "Non-Disturbing Optical Power Monitor for Links in the Visible Spectrum Using Polymer Optical Fibre", Measurement Science and Technology, 26, 085201 (7pp), 2015.
3. Velickov Engineering, www.velickov.eu/pof_en.html1#pof_fpm.
4. S. Feistner, H. Lichotka and H. Poisel, "Thermal stability of POF couplers", 10th International Conference on Plastic Optical Fibers (ICPOF 2001), Amsterdam, The Netherlands, pp. 251-256, September 27-30, 2001.
5. R.M. Ribeiro and M.M. Werneck, "Improvements of resolution and precision of a wavelength-encoded electrical current sensor using an ultra-bright light-emitting diode transducer", Transaction of the Institute of Measurement and Control, 30, pp. 153-171, 2008.
6. O. Ziemann, J. Krauser, P.E. Zamzow and W. Daum, "POF Handbook: Optical Short Range Transmission Systems", 2nd edition, Springer-Verlag, Berlin, Germany, 2008.

7. T.A.M.G. Freitas, R.D. Oliveira, V.H. Silva, V.N.H. Silva, A.P.L. Barbero, C.B.M.P. Leme and R.M. Ribeiro, "The enhancement of the performance of a simple optical power monitor for PMMA plastic optical fibre links", 24th International Conference on Plastic Optical Fibers (ICPOF 2015), Nuremberg, Germany, September 22-24, 2015.
8. T.A.M.G. Freitas, R.D. Oliveira, V.H. Silva, V.N.H. Silva, A.P.L. Barbero, C.B.M.P. Leme and R.M. Ribeiro, "An optical power monitor for 2-channels WDM links using PMMA plastic optical fibers", 24th International Conference on Plastic Optical Fibers (ICPOF 2015), Nuremberg, Germany, September 22-24, 2015.
9. C.-A. Bunge, R. Kruglov and H. Poisel, "Rayleigh and Mie scattering in polymer optical fibers", Journal of Lightwave Technology, 24, 8, pp. 3137-3146, 2006.
10. H. Poisel, G. de Preux, A. Bachmann, O. Ziemann and K.-F. Klein, "Characterisation of side-emitting fibers", 18th International Conference on Plastic Optical Fibers (ICPOF 2009), paper PS_40, Sydney, Australia, September, 9-11, 2009.
11. J. Fischer, H. Poisel, A. Bachmann, F. Süß, A. Wagner and K.F. Klein, "A standard proposal for characterizing side-emitting fibers", 20th International Conference on Plastic Optical Fibers (ICPOF 2011), paper POS_051, Bilbao, Spain, September 14-16, 2011.
12. O. Ziemann, J. Krauser, P.E. Zamzow and W. Daum, "POF Handbook: Optical Short Range Transmission Systems", 2nd edition, pp. 52, 53, 113 and 116, Springer-Verlag, Berlin, Germany, 2008.
13. D. Gloge, "Optical power flow in multimode fibers", Bell System Tech. Journal, 51, pp. 1767-1783, 1972.

V

Biographies

List of Biographies

Regina Célia da Silva Barros Allil was born in Rio de Janeiro, Brazil. She received her BSc Degree in electronic engineering from the Faculdade Nuno Lisboa, Rio de Janeiro, in 1988, and the M.Sc. degree from the Biomedical Engineering Program, Federal University of Rio de Janeiro (UFRJ), Brazil, in 2004. Her D.Sc. degree was obtained from the Electronic Engineering Program, Instrumentation and Photonics Laboratory, UFRJ, in 2010. She is currently a researcher with the Brazilian Army Technological Center (CTEx), Rio de Janeiro and a researcher at Instrumentation and Photonics Laboratory, UFRJ. Her research interest lies in fiber optics sensors, optoelectronic instrumentation and biosensors.

Jon Arrúe received the M.Sc. Degree in electronic physics, completed a 12-month postgraduate course in electronics and a 12-month postgraduate course in telecommunications, and received the Ph.D. degree in optical fibers from the University of the Basque Country UPV/EHU in Bilbao, Spain, in 1990, 1991, 1992, and 2001, respectively. He is currently a Professor with the Department of Electronics and Telecommunications in the Faculty of Engineering of Bilbao, UPV/EHU, and is also involved in international research projects with other universities and companies. Dr. Arrue was a recipient of a special award for his thesis and a European acknowledgement of the Ph. D. degree.

Igor Ayesta received the M.Sc. Degree in electronic engineering and the Ph.D. degree in optical fibers from the University of the Basque Country UPV/EHU, Bilbao, Spain, in 2008 and 2013, respectively. He is currently a Lecturer with the Department of Applied Mathematics, School of Engineering of Bilbao, UPV/EHU, and is involved in research projects in collaboration with universities and companies from Spain and other countries in the field of polymer optical fibers. Dr. Ayesta was a recipient of European acknowledgement of the Ph. D. degree.

Thomas Becker was born in Nuremberg, Germany, in 1982. He received the Dipl.-Ing. (FH) degree in electrical engineering from Technische Hochschule Nürnberg Georg Simon Ohm in 2008. Between 2007 and 2013 he was with Siemens AG, Fürth, Germany as a Software Developer and Project Manager. In 2013 he received the Master of Science degree in electrical engineering (with honors) from University of Hagen before he joined the Polymer Optical Fiber Application Center in Nuremberg, where he is currently working towards his PhD degree. His interests lie in the field of polymer optical fiber modelling, which includes ray tracing simulations and analytical approaches.

Markus Beckers was born on 1st April 1982 in Würselen, Germany. Via his contact to Prof. Gries, who supervised his membership of the German National Academic Foundation since 2011, he got in first touch with the ITA. So, since October 2011, he heads the research group POF in the department for Chemical Fibers of the Institut für Textiltechnik (ITA) at the RWTH Aachen University. The research is focused on melt spinning of SI- and GI-POF, plus its simulation.

Christian-Alexander Bunge was born in Berlin, Germany, in 1973. He received the Dipl.-Ing. degree in electrical engineering from university of technology Berlin (TU Berlin) in 1999. Until 2002, he was with TU Berlin's photonics group of Prof. Petermann, where he received his PhD. From 2002 to 2004, he was with the Polymer Optical Fiber Application Center (POF-AC) in Nuremberg, where he was responsible for fiber modeling and short-haul systems design. In 2004, he joined the TU Berlin as senior scientist, working on high- speed optical transmission systems, studying nonlinear optics, and modeling of optical components. In 2009, he became professor at Deutsche Telekom's university for telecommunication in Leipzig, where his main research interests are short-haul and access transmission techniques, multimode glass and polymer-fiber links, signal processing and optical sensors.

Cesar Cosenza de Carvalho was born in Rio de Janeiro, state of Rio de Janeiro, Brazil, in 1966. He received B.Sc., M.Sc. and D.Sc. degree in electronic engineering from the Federal University of Rio de Janeiro (UFRJ - Brazil), in 1989, 1994, and 2000, respectively. Currently, he is a researcher in the Instrumentation and Photonics Laboratory (LIF) at the Electrical Engineering Program of UFRJ. His research interests include fiber optics, sensors, transducers, and optoelectronic instrumentation.

Branko Drljača was born in Kragujevac, Serbia, in 1981. He received the B.E. degree in physics from the Faculty of Natural Science, University of Kragujevac, Kragujevac, Serbia, in 2006, and Ph.D. degree in physics from the Faculty of Natural Science, University of Kragujevac, Kragujevac, Serbia, in 2011. In 2009, he joined the Department of Physics, University of Priština, Kosovska Mitrovica, Serbia as Assistant, and in 2012. became Assistant Professor. His current research interests include optical fibers, transfer characteristics of optical fibers, diffusion processes. He is member of steering committee of Serbian Physical Society since 2013.

Jakob Fischer received the M.Sc. degree in Applied Research in Engineering Sciences from the "Ohm-Hochschule", Nuremberg, Germany in 2012. At the present he is working at the Polymer Optical Fiber Application Center at the "TH-Nuremberg". There he is specialized in optical fiber metrology, optical energy transmission, sensor technologies and side emitting fibers.

Martin Gehrke was born in Nuremberg, Germany, in 1988. He received the Bachelor of Engineering degree in electrical engineering in 2011 and subsequently the Master of Engineering degree in mechanical and electrical systems in 2013 from Technische Hochschule Nürnberg Georg Simon Ohm. Since 2013 he is with the Polymer Optical Fiber Application Center in Nuremberg, where he is currently working towards his PhD degree. His interests lie in the field of fiber modelling and simulation, especially regarding backscatter simulation estimations and approaches.

Thomas Gries is professor at the RWTH Aachen University for textile machinery and director of the Institut für Textiltechnik (ITA) of RWTH Aachen University. He is author of more than 750 scientific papers as well as more than 160 presentations. Since 2013 he is Honorary professor at Lomonosov Moscow State University, Moscow, Russian Federation.

Daniel Grothe was born on 13th September 1992 in Hamburg, Germany. He started studying mechanical engineering in 2012 at the RWTH Aachen University. Since 2015 he has been working at the Institut für Textiltechnik (ITA) of the RWTH Aachen University as a student employee. With Markus Beckers and Prof. Christian-Alexander Bunge he works on the simulation of amorphous polymers in the context of melt spinning of SI- and GI-POF.

M. A. Illarramendi received the M.Sc. degree in Solid-State Physics from the University Autónoma de Madrid and the Ph.D. degree in Physics from the University of the Basque Country (UPV/EHU) getting the Ph.D. Excellence Award (1991). She is currently Professor in the area of Physics and Electromagnetism at the Department of Applied Physics I at the UPV/EHU. Her current research activities studies are related with propagation, generation and amplification of light in Polymer Optical Fibers.

Felipe Jiménez received the M.Sc. degree and the Ph.D. degree in Industrial Engineering from the University of the Basque Country UPV/EHU in Bilbao, Spain, in 1990 and 2000 respectively. He is currently an Associate Professor with the Department of Applied Mathematics in the Faculty of Engineering of Bilbao, teaching Numerical Analysis, Algebra and Statistics. His research activity is carried out in the Applied Photonics Group of UPV/EHU, where his main contributions specialize on the numerical simulation of light propagation in passive and active plastic optical fibers.

Roman Kruglov was born in Seversk, Russia, in 1982. He received the degree of engineer in 2004 and Ph.D degree in 2007, both from Tomsk State University of Control Systems and Radioelectronics, Russia. Since 2009, he has been with the Polymer Optical Fiber Application Center (POF-AC) at Technische Hochschule Nürnberg Georg Simon Ohm, working on optical transmission systems and digital signal processing.

Sven Loquai was born in Coburg, Germany, in 1980. He received the undergraduate degree from TH Nürnberg Georg-Simon-Ohm und postgraduate degree from La Trobe University in Melbourne. In 2014 he received the Ph.D. degree from the Friedrich-Alexander-University Erlangen-Nürnberg. His interests are in the field of high-speed POF transmission systems, semiconductor devices, optical fiber telecommunication and RF/microwave designs.

Birgit Lustermann Education Sep 1986 – Feb 1991: University of Technology, Ilmenau. Graduate (Diploma) Engineer of Electrical Engineering course: Technical Cybernetics and Automation Technology with specialization in Process Measurement and Sensor Technology. Oct 2015: Dissertation "Modeling of tubular optical fibers using the example of an optical-electrical combination conductor", University of Technology Ilmenau, Faculty of Mechanical Engineering.
Experience Apr 1991- Oct 1999: trainer for courses in electrical engineering at GfM (Society for micro electronics) Göttingen GmbH. Since1999 Laboratory engineer at the University of Applied Sciences Nordhausen, Faculty of Engineering.

Alicia López received the M.Sc. degree in Telecommunications Engineering and Ph.D. degree from the University of Zaragoza (UZ), Zaragoza, Spain, in 2002 and 2009, respectively. In 2002, she joined the Photonic Technologies Group (GTF), Aragon Institute of Engineering Research (i3A). Since 2004, she has been Assistant Professor in the Departamento de Ingeniería Electrónica y Comunicaciones, UZ. Her research interests include the use of plastic optical fibers in communications applications and the design and evaluation of optical networks.

M. A. Losada received her Ph.D. in Physics from the Universidad Complutense de Madrid (Spain) in 1990. From 1991 to 1994 she was working as a postdoctoral researcher at the McGill Vision Research Laboratories of McGill University in Montréal (Canada), and afterwards, in the Institute for Research in Optics of the Scientific Research Council of Spain in Madrid (Spain). In 1997, she joined the Photonic Technologies Group (GTF) in the Aragon Institute for Engineering Research (i3A) of the Universidad de Zaragoza (Spain) where she has a tenure position as an associate professor. Her research interests are centered in optical communications based on plastic optical fibers and in optical networks.

Michael Luber was born in Fuerth, Germany, in 1976. He received the Dipl.-Ing. (FH) degree in electrical engineering from Technische Hochschule Nürnberg Georg Simon Ohm in 2002. Since 2002, he has been with the Polymer Optical Fiber Application Center in Nuremberg, working on characterization of polymer optical fibers and measurements under environmental conditions.

Fernando Luiz Maciel was born in Rio de Janeiro in 1974, and graduated in electronic engineering from Universidade Federal do Rio de Janeiro (2006), with emphasis on instrumentation, control systems, and correction and protection of electrical power systems. He currently works with Research and Development at the Photonics and Instrumentation Laboratory – Universidade Federal do Rio de Janeiro, in Brazil.

Eberhard Manske holds a professorship "Production and Precision Measurement Technology" at the Ilmenau University of Technology. He was spokesman of the collaborative research centre "SFB 622 – Nanopositioning and Nanomeasuring Machines" from 2008 to 2013. The research activities are continued by a new Competence Centre Nanopositioning and Nanomeasuring Machines.

After the completion of his studies of electrical engineering in 1982, Prof. Manske worked as a scientific co-worker at the Institute of Process Measurement and Sensor Technology at the Ilmenau University of Technology. In 1986, he obtained the doctoral degree, and in 2006 the postdoctoral lecturing qualification. He focuses his research work mainly on nanopositioning and nanomeasuring technology, fibre-coupled laser interferometry, laser stabilization, optical and tactile precision sensors and scanning probe techniques.

Claudia Barucke Marcondes has Graduation at Electrical Engineering from Pontifical Catholic University of Rio de Janeiro PUC-RJ (1997), Master degree at Electric Engineering from Pontifical Catholic University of Rio de Janeiro PUC-RJ (2000) and DSc at Electric Engineering from Pontifical Catholic University of Rio de Janeiro PUCRJ (2009). Pos-Doctoral in Microwave System at CETUC (Telecommunication Center at Pontifical Catholic University of Rio de Janeiro) - 2013/2014, and Pos-Doctoral in Optical System at LaCOp (Optical Comunnications Lab at UFF - Federal University) - 2015/2016. Has experience in Electric Engineering, on the following subjects: optical and microwave systems, optical networkacting, Telecommunications Management Network, microwave photonics, NGN - next generation network, optical ethernet, RF subcarriers, and RFID.

Javier Mateo received the M.Sc. degree in Electrical Engineering from the Polytechnic University of Madrid (UPM), Madrid, Spain, and the Ph.D. degree from the University of Zaragoza (UZ), Zaragoza, Spain, in 1989 and 2000, respectively. From 1989 to 1993, he was with Cables de Comunicaciones S. A., Zaragoza, Spain, where he worked on fiber optic sensors and optical communications. In 1993, he joined the Electronic Engineering and Communications Department of the University of Zaragoza, where he is currently Associate Professor of Optical Fiber Communications. His professional research interests are in signal processing, in particular, applied to biomedical signals, fiber optic sensors, and optical communication systems. He is part of the Aragon Institute of Engineering Research (i3A).

Fábio Vieira Batista de Nazaré was born in Maceió, Estate of Alagoas, Brazil, in 1984. He received the Degreein electronic engineering from the Universidade Federal de Pernambuco, Recife, Brazil, in 2007. He was a Research Engineer with the Nuclear Instrumentation Laboratory, Regional Center of Nuclear Sciences, Recife. He received the M.Sc. degree in electrical engineering from the Institute for Post-Graduate Studies and Research in Engineering, Federal University of Rio de Janeiro, Rio de Janeiro, Brazil, in 2010, where he also received the D.Sc. degree at the Instrumentation and Photonics Laboratory, Electrical Engineering Program, in 2014. Currently, he works as an Electronic Engineer mainly developing projects concerning optical sensors and fiber Bragg grating devices.

Hans Poisel received his M.Sc. in 1977 in physics at the Technical University Munich and his PhD in 1983 at the Faculty of Electrophysics. In 1985 he joined MBB in Ottobrunn / Munich, active in development of fiber optic sensors, leading the fiber optic gyro project and furthermore coordinating all fiber optic activities of the company as a whole. Since 1991 he has been professor for Technical Optics and Optical Communication at the University of Applied Sciences Nuremberg, Germany. He published more than 100 contributions devoted to optical properties of Plastic Optical Fibers and their applications in international conferences and journals and holds more than 50 patents in the area of fiber optics. Currently he is a member of the International Committee for Plastic Optical Fibers (ICPOF) and director of Polymer Optical Fiber Application Center (POF-AC) in Nuernberg. He is a member of OSA, IEEE, VDE and DPG.

Ricardo M. Ribeiro has BSc. in Physics (1985), MSc. in Physics (1989) and DSc. in Physics (1995) on experimental Photonics applied to Telecommunications, PUC-Rio. From 1996 to 2005 he worked as Researcher on photorefractive fibres, external cavity mode-locked lasers (British Telecom Research Laboratories - UK) and optoelectronics applied to instrumentation and sensing. Since 2006 he is a permanent Professor (Associate Professor I in 2015) of applied Electromagnetism and Photonics Researcher of CNPq on the Laboratório de Comunicações Ópticas (LaCOp) of the Departamento de Engenharia de Telecomunicações of Universidade Federal Fluminense. He has more than 80 internationally published papers, 8 filed patents (+1 in preparation) and many technological developments. From Aug/2012 to Nov/2013 he joined the Optics Department of Telecom Bretagne in Brest/France for a Post-Doc stage working with "All-Photonic Digitising of Radio-over-Fibre Signals". His main interests and R&D efforts are on Integrated Optic and Fibre-Optic Devices for Telecommunications, all-optical signal processing, Microwave-Photonics (Radio-over-Fibre technology), Polymer Optical Fibres for Datacom, opto-acoustic devices, optical amplifiers, multimode transmission systems and Fibre-to-the-X technologies.

Daniel Moreira dos Santos was born in Rio de Janeiro, Brazil. He graduated as electronic engineer from Federal Centre of Technology Education Celso Suckow da Fonseca, Rio de Janeiro in 2010. Currently, He is an M.Sc student at Electric Engineering Program in the Photonics and Instrumentation Laboratory (LIF) at the Universidade Federal do Rio de Janeiro. Also, he is a researcher at Institute for Postgraduate Studies and Research in Engineering – COPPE/UFRJ. His research interests include sensors, instrumentation and fiber optics.

Paulo Acioly M. dos Santos received his PhD in physics from the State University of Campinas, Brazil, in 1989, working with the optical properties and applications of the photorefractive effect in BSO-type crystals. Dr. Santos is currently a research physicist and head of the Laboratório de Óptica não Linear & Aplicada, associated professor and supervisor of undergraduate and graduate students at the University Federal Fluminense, also in Brazil. He is currently interested in studies and applications using the photorefraction, digital holography and related phenomena, also in interferometry and digital image processing, and in all this tecniques applied to material science and mechanical engineering.

Bernhard Schmauss received the Dipl. Ing. and Dr. Ing. degrees in electrical engineering from the University Erlangen-Nuremberg in 1989 and 1995, respectively. In 1995, he joined Lucent Technologies, in Nuremberg, Germany. From 2003 to 2005 he was professor at the electrical engineering department at the university of applied sciences in Regensburg, Germany. Since October 2005 he is professor for optical high frequency technology and photonics at the institute of microwaves and photonics of the University Erlangen-Nuremberg. His current research interests are fiber lasers, medical application of photonics, optical sensors, few mode fiber propagation, optical transmission systems, advanced modulation formats, optical signal regeneration and electronic distortion mitigation. He is principal investigator of the "Erlangen Graduate School in Advanced Optical Technologies" and chairman of the elite study program "Master in Advanced Optical Technologies".

Vinicius N. Henrique Silva has BSc in Telecommunication Engineering (2006), MSc. in Optics Communication (2009), Ph.D. in Physics/Engineering (2013) on Liquid Crystals applied to stereoscopic applications, Télécom-Bretagne (France). Since 2013 he is a permanent Professor (Adjunct Professor I) at Universidade Federal Fluminense working in the optical communication laboratory (LACOP). His main R&D area are optic devices for datacom and sensor applications, atmospheric turbulence applied to free-space optics, polymer optical fibres and optical access network.

Pavol Stajanca holds a Master's degree in "Optics, lasers and optical spectroscopy" from Comenius University in Bratislava. Previously he has worked as a researcher at International Laser Centre in Bratislava, where he focused on ultrafast nonlinear optics, all-optical signal processing, photonic crystal fibers and fiber-based laser devices. Since 2014, he has been with BAM Federal Institute for Materials Research and Testing (Berlin, Germany) as a part of EU ITN project TRIPOD, that is located in the field of polymer optical fiber sensors with particular focus on the fiber Bragg grating technology.

M. Viehmann Professur Industrieelektronik Hochschule Nordhausen; Abgeschlossene Qualifizierungen: Studium Industrielle Elektronik (FH), Studium Elektrische Energietechnik (UNI), Studium Pädagogik (UNI), Promotion an Universität Magdeburg; Forschungsschwerpunkte: Diagnosesysteme, Energiemanagementsysteme, Störlichtbogenschutz

Marcelo Martins Werneck was born in Petrópolis, Brazil. He received the Degree in electronic engineering from the Pontifícia Universidade Católica of Rio de Janeiro, Rio de

Janeiro, Brazil, in 1975 and his M.Sc. degree from the Biomedical Engineering Program, Federal University of Rio de Janeiro (UFRJ), Rio de Janeiro, in 1977. His Ph.D. degree was obtained from the University of Sussex, Brighton, U.K., in 1985. He has been with UFRJ since 1978, where he is currently a Lecturer and Researcher. He is also the Coordinator of the Instrumentation and Photonics Laboratory at the Electrical Engineering Program, UFRJ. His research interests include fiber optics sensors, nanobiosensors, transducers and instrumentation.

Olaf Ziemann studied physics at the der Karl-Marx-University Leipzig. He made his PhD at the Technical University Ilmenau on the topic optical coherent detection. Between 1995 and 2001 he worked on the Technologiezentrum of the Deutschen Telekom in Berlin, before he took over the scientific leadership of the Polymer Optical Fiber Application Center (POF-AC) at the technical university of applied sciences Nuremberg. He is active in the Information Technology Society (ITG) and published the "POF Handbook" in 2007.

Joseba Zubia received the M.Sc. Degree in solid-state physics and the Ph.D. degree in physics from the University of the Basque Country UPV/EHU in Bilbao, Spain, in 1988 and 1993, respectively. His Ph.D. work focused on the optical properties of ferroelectric liquid crystals. He is currently a Full Professor with the Department of Engineering of Communications, School of Engineering of Bilbao, University of the Basque Country. He has more than 22 years of experience doing basic and applied research in the field of polymer optical fibers and is currently involved in research projects in collaboration with universities and companies from Spain and other countries in the field of polymer optical fibers and fiber-optic sensors. Prof. Zubia was a recipient of a special award for Best Thesis in 1995 and the Euskoiker award in 2009.

www.ingramcontent.com/pod-product-compliance
Lightning Source LLC
Chambersburg PA
CBHW082205220526
45470CB00010B/3054